ON THE TRAIL OF
STARDUST

The Guide to Finding Micrometeorites:
Tools, Techniques, and Identification

Jon Larsen

VOYAGEUR
PRESS

Inspiring | Educating | Creating | Entertaining

Brimming with creative inspiration, how-to projects, and useful information to enrich your everyday life, Quarto Knows is a favorite destination for those pursuing their interests and passions. Visit our site and dig deeper with our books into your area of interest: Quarto Creates, Quarto Cooks, Quarto Homes, Quarto Lives, Quarto Drives, Quarto Explores, Quarto Gifts, or Quarto Kids.

First published in 2019 by Voyageur Press, an imprint of The Quarto Group,
100 Cummings Center Suite 265D, Beverly, MA 01915 USA.
T (978) 282-9590 F (978) 283-2742 QuartoKnows.com

Voyageur Press titles are also available at discount for retail, wholesale, promotional, and bulk purchase. For details, contact the Special Sales Manager by email at specialsales@quarto.com or by mail at The Quarto Group, Attn: Special Sales Manager, 100 Cummings Center Suite 265D, Beverly, MA 01915 USA

10 9 8 7 6 5 4 3 2 1

ISBN: 978-0-7603-6458-1

Digital edition published in 2019
eISBN: 978-0-7603-6459-8

Library of Congress Cataloging-in-Publication Data

Names: Larsen, Jon, 1959- author.
Title: On the trail of stardust : the guide to finding micrometeorites : tools, techniques, and identification / by Jon Larsen.
Other titles: On the trail of star dust | Guide to finding micrometeorites
Description: Minneapolis, Minnesota : Voyageur Press, an imprint of The Quarto Group, 2019. | Includes index.
Identifiers: LCCN 2018051038 | ISBN 9780760364581 (pbk.)
Subjects: LCSH: Meteorites--Identification. | Rocks--Identification.
Classification: LCC QB755 .L3725 2019 | DDC 523.5/1--dc23
LC record available at https://lccn.loc.gov/2018051038

ACQUIRING EDITOR: Dennis Pernu
PROJECT MANAGER: Alyssa Bluhm
ART DIRECTOR AND COVER DESIGN: Cindy Samargia Laun
PAGE DESIGN AND LAYOUT: Laura Shaw Design
FRONT COVER IMAGE: Yuri Beletsky (Creative Commons)

Printed in China

MIX
Paper from
responsible sources
FSC® C016973

CONTENTS

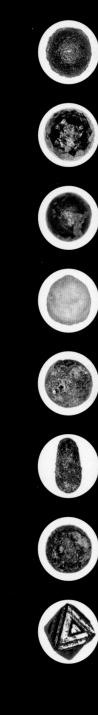

INTRODUCTION

AN UNEXPECTED DISCOVERY

It is the oldest matter there is. Nothing has traveled farther. It is the building blocks of galaxies, planets, and even humans. We are all made of stardust.

For more than a century, scientists have searched for stardust—the enigmatic micrometeorites—but found them only in extremely clean and remote areas, such as the Antarctic blue ice or, more recently, in space. Many attempts have been made to find these small mineral particles in populated areas, but a confusing array of similar-looking manmade objects has been an unsurmountable obstacle. The consensus was that it is simply not possible to pick out *one* micrometeorite among zillions of other particles.

After an incident in 2009, when a little black dot from above literally landed on my porch table, I wanted to find out more about micrometeorites. I was intrigued by the contradiction between the common belief that they could not be found and the known global influx rate of micrometeorites measured by radar. Before I knew it, I had become a full-time stardust hunter.

To find micrometeorites and separate them from terrestrial dust, we must know what to look for and what to disregard. At the time I began to search, the published pictures of micrometeorites from the Antarctic were mainly black-and-white scanning electron microscope (SEM) section images, which are poor representations of what micrometeorites really look like. And with regard to terrestrial imposters, there was a plentitude of speculation but very little empirical data. Citizen scientists had searched for micrometeorites in roof gutters, but none of these efforts had resulted in verification of a single extraterrestrial particle.

In the spring of 2010 I started systematic research on dust samples from populated areas. Initially I looked at skyward-facing hard surfaces where particles could accumulate over time, such as roads, parking lots, and industrial areas. Then I graduated to other cities, countries, mountains, beaches, deserts—everywhere. A decade later, I can look back upon one thousand field searches in more than fifty countries on six continents.

Stardust in the sky: The Rosette Nebula is an expanding dust cloud and raw material for new worlds. *Jan Inge Berentsen Anvik*

The samples are examined in a Zeiss binocular microscope. Interesting particles are picked out, photographed with a USB microscope, and archived. I have established a photo database (now containing photos of more than forty thousand individual objects), kept an illustrated journal, and tried to find patterns in the field samples. All the while, I put my complete trust in pure empiricism and established the Facebook page Project Stardust to share the results. Today this is the main website for micrometeorite hunters all over the world.

In the beginning the diversity of particles in the dust seemed infinite and chaotic, but with time I started to recognize the most common ones. Micrometeorites are rare and evenly distributed, so an abundance of one type of particle in the dust of one area is one indication of terrestrial origin. At the early stages of my research I contacted the handful of professional scientists in the micrometeorite world, and they all agreed.

The breakthrough came in February 2015, when Matthew Genge from London's Imperial College verified my first urban micrometeorite: a barred olivine beauty with dendritic magnetite crystals sprinkled all over its surface. The stone was only 0.27 millimeter in diameter, but at last I knew what to look for. I immediately started to search for similar stones and found them. Within my first season I had a collection of more than five hundred pristine micrometeorites that included all of the most common types.

After the release of my book *In Search of Stardust*, now available in several translations, there are stardust hunters in many countries, and they have found amazing space rocks. This has opened a new branch of citizen science, which at the same time is cutting-edge scientific research. It turns out that we are surrounded by stardust, and it can be found everywhere. All you need is a magnet, a sieve, and magnification, the method for which is explained for the first time in this book. The key, however, is to know what to look for and what to disregard. Micrometeorites look like nothing else down here on Earth.

To better photograph micrometeorites, Jan Braly Kihle and I have constructed a photo rack based on a modified Olympus camera with new and prototype components (both hardware and software), which have resulted in the high-resolution color photos in this book. Studying the morphological details in high-resolution color is crucial to understanding what to search for in field samples. Today the development of electronic photo equipment is exploding, and there is reason to assume that new and improved solutions for photographing micrometeorites will find their way to the market.

Most of the cosmic spherules retrieved on Earth have a *chondritic* chemistry, which is one of the reliable criteria in micrometeorite verification. When a micrometeoroid enters Earth's atmosphere at a steep angle, it goes through a rapid and unique transformation: melting, differentiation, and recrystallization

Until recently, micrometeorites have been found mainly in the ultraclean environment of the Antarctic. Here, a Korean scientist takes snow samples in search of stardust. The same particles are in the nearest rain gutter. *Lee Jong-ik, Polar Research Institute, South Korea (Creative Commons)*

in hyperspeed, resulting in an aerodynamic stone. These forms, together with characteristic surface textures, are in most cases sufficient to visually identify a micrometeorite. This would have been considered science fiction just a few years ago.

This book is all about showing the methodology to help you find your own micrometeorites. Since the discovery of the first urban micrometeorite in 2015, this methodology has improved continuously. Today, the results are up to ten thousand times better than they were in 2015, matching those earlier Antarctic findings. Nevertheless, we are still in the beginnings of this new field. I recommend reading this book to get started and then using your imagination to improve the results.

How to separate alien rocks from their terrestrial imposters is shown here for the first time, step by step, illustrated and explained in an easy way. The first part of the book is about gear and equipment—how to use it, and where and when. Then comes a brief presentation on what you can expect to find: industrial and naturally occurring particles as well as micrometeorites. Use this part of the book as a roadmap through the maze of contaminants found in dust from populated areas.

Now *you* can find these fantastic space rocks on the nearest rooftop—happy hunting!

Project Stardust, *Jon Larsen*
University of Oslo (UiO) / project.stardust@getmail.no

1 MICROMETEORITES EVERYWHERE

MICROMETEORITES belong to the oldest matter there is. They are mineral remnants from before the planets were formed and may even contain particles older than the sun that have traveled farther than anything else on Earth. We are just beginning to explore these microscopic alien stones, yet they are everywhere.

WHAT IS A SPHERULE?

Most micrometeorites retrieved on Earth are cosmic *spherules*. A spherule is a rounded object, a solidified melt droplet of stone and/or metal. The sphere is nature's efficient solution for maximum volume and minimum surface. Surface tension while the micrometeorite is still in a liquid state bends the object into a sphere. Raindrops are also formed this way.

In nature, rock can melt and create spherules in three ways: by volcano, lightning, and meteorite. On an uninhabited world, we would only have to distinguish these three types of naturally occurring spherules from one another to identify the extraterrestrial particles. There are, however, traces of human activity everywhere on Earth, in all sediments and layers younger than the Industrial Revolution dating to the 1760s. This is why scientists at the Scott-Amundsen Base at the South Pole drilled down through the ice to the strata from one thousand years ago to melt drinking water. There, however, they found other types of impurities in the ice: micrometeorites.

Power tools and industrial processes produce a huge number of *anthropogenic* (manmade) spherules, which are everywhere on Earth's surface. The search for stardust is therefore a question of separating the terrestrial (both natural and manmade) from the extraterrestrial. When you know what micrometeorites look like, and what to disregard as anthropogenic spherules, you can find stardust. But where shall we begin the search?

Stardust falling on Earth. An exchange of small mineral particles is occurring across the universe, both inside and outside our solar system. *Shutterstock/Dream Ideas*

SEARCHING FOR THE RIGHT PLACE

To find micrometeorites, we must understand two concepts. The first is *accumulation*, the gradual increase in number or amount over time. Despite the total global influx of approximately 100 metric tons of cosmic dust particles *per day*, the rate on the ground is low, simply because there's so much ground to cover, not to mention that water covers more than 70 percent of Earth's surface. We can expect one extraterrestrial particle with a diameter of 0.1 millimeter per square meter per year.

The longer a skyward surface has accumulated particles from above, the more micrometeorites there are to be found. Contrary to existing literature on the subject, cosmic spherules do not erode "in weeks." In my studies I have found that the older an accumulating area is, the more micrometeorites one is likely to find. The exception is if prevailing winds are especially unfavorable, in which case the yield might be low. Of nearly two thousand cosmic spherules examined, I have seen signs of slight erosion on only a couple of particles. Our hunt for stardust, therefore, begins with a search for a hunting ground with favorable accumulation.

The other concept we must understand is *signal-to-noise ratio*. In the context of micrometeorites, this means how many micrometeorites (signal) there are compared to terrestrial particles (noise).

For years, micrometeorite hunters have built traps of various types. Few have succeeded because cosmic dust particles are rare. To catch hundreds of cosmic spherules, a trap would have to be the size of a football field and accumulate particles over decades. The challenges connected with such a construction have discouraged more than one good scientist. There are, however, "traps" already in place and ripe for harvesting: roofs.

When I started to hunt for micrometeorites, I began by searching for large, old accumulating areas, such as roads and parking lots. But I did not find anything other than myriad anthropogenic spherules. Not until I moved the search up one floor closer to the sky—to the roof's rain gutter—did I start to find extraterrestrial treasures. The explanation is an improved signal-to-noise ratio.

On the roof there is less human activity, and thus there are fewer anthropogenic spherules. The difference between the signal-to-noise ratio in roof dust and urban road dust is enough for us to find micrometeorites in the former while making the latter very difficult. Consequently, the hunt for urban micrometeorites begins with finding the right roof. Thanks to the stronger signal-to-noise ratio, almost any roof or surface above the turbulent ground will do. A pitched roof with an easily accessible gutter is an excellent place to start. The larger and

On a flat roof, most particles accumulate around drains. Use a stiff dishwashing brush to loosen the compact sedimentation, and then extract the magnetic content.

older a roof is, the better. Particles fall from above, roll down roof tiles or shingles, and accumulate in rain gutters. By placing a bucket under the downspout for a year or two, one may even catch those that the rainwater washes away.

The fewer particles the roof decking contributes to the content of the gutter, the better. Glazed roof tiles, metal plates, stone, wood, and vinyl are favorable roof materials for a successful micrometeorite hunt. With experience you will recognize a promising roof by its accumulating properties and signal-to-noise ratio. Then the result of the hunt accelerates.

Pitched roofs aren't the only place to search. Flat roofs, which usually have safety walls around the edges, serve as micrometeorite traps. On roofs such as this, loose particles accumulate at the lowest points: around the drains, along the walls, and in the corners.

A flat roof is not entirely flat but constructed of slightly pitched sections, each of which leads water into a drain. Drains are usually located in the middle of the roof, making it a bit safer for us to conduct our search away from roof edges. Moreover, the drain itself is often elevated slightly above the lowest point, leaving the small particles to accumulate there for star hunters such as us. Crush or pulverize any lumps and extract the magnetic particles.

It is important to emphasize HSE (health, safety, and environment). Remember your sunscreen. Also, roofs are the domain of birds, and in some places their droppings may be the main constituents of the roof sample. This can present a biohazard. I always use gloves; a dust mask (N95 type) is also recommended. A disinfecting soak of the sample in rubbing alcohol prior to the rinse will make it even safer to handle.

Micrometeorites are so small that the prevailing winds, not gravity, ultimately determine where they end up on a flat surface. In this respect cosmic dust may be considered a type of *aeolian* sediment (i.e., carried by the wind). Safety walls along the edges of a flat roof or a new floor rising opposite to where the prevailing winds come from (see photo on page 16) will act as micrometeorite traps, allowing the extraterrestrial particles to accumulate. If, in addition to these factors, the roof is covered with PVC vinyl, the signal-to-noise ratio will be stronger because PVC does not create dust particles that can be confused with micrometeorites, offering an optimal starting point for a successful hunt. Use a dust broom to gather loose debris and extract the magnetic particles. When finished, take a few photos of the roof and the sample in situ.

In a place where you suspect micrometeorites are abundant, it is recommended that you also take a nonmagnetic sample. As mentioned, the micrometeorite collection from the water well at the South Pole base contains around 20 percent nonmagnetic glass spherules—completely melted micrometeorites. To search for these beautiful stones in a promising place where the signal-to-noise ratio is expected to be particularly strong, we can scrape up the remaining particles after having extracted the magnetic ones. Make sure to mark the sample "nonmagnetic."

The rooftop finds in urban micrometeorite collections are found on the roofs of buildings no older than 50 years. Consequently, it can be assumed that these stones have a terrestrial age of 0 to 50 years, which makes them fresh compared to most micrometeorites in other collections. Most of the Antarctic stones, for example, have a terrestrial age of 1,000 to 1,000,000 years and are weathered accordingly. Some are eroded beyond recognition. Therefore, the urban micro-

A large roof covered with PVC vinyl. The magnetic sample was around 30g, but after rinsing and screening for size (150–400µm) it was down to 9.3g, containing a record of 164 micrometeorites!

A successful hunt for stardust often starts with a search for the right roof. With Google Earth it is possible to search an area for the most promising places. A large, flat roof with safety walls along the edges may function as a giant micrometeorite trap. The light roofs covered with PVC do not contribute to the dust content on the roof and in most cases are the best hunting grounds for micrometeorites.

meteorites look a bit different—the pristine cosmic spherules are less eroded and have distinct surface textures. This enables us to identify fresh micrometeorites by visual examination.

By monitoring a skyward-facing area at regular intervals, it should be possible to be even more precise in future sampling, perhaps down to the week—or even day—that a micrometeorite fell to Earth. With careful preparation and cleaning of the collection area around annually reoccurring meteor showers, it should be possible to identify material from some of the comets and possibly detect variations in the influx rate over time.

We are still in the beginning stages of understanding micrometeorites. We have managed to break the code and find micrometeorites on roofs, where the signal-to-noise ratio is better than it is on the ground below. But in the future, we may even be able to separate the trillions of micrometeorites from the vast amounts of urban road dust.

Micrometeorites fall like a gentle cosmic rain in equal amounts everywhere on Earth. There is no need for an expensive expedition to find them—the nearest roof will do. To find *a lot* of micrometeorites, however, it is possible to use Google Earth to search for exceptionally large roofs. Schools, shopping malls, sports arenas, airports, and industrial buildings are superb hunting grounds. The best results are found on large, old, PVC-covered flat roofs with security walls around the edges. In such a place one may find a hundred micrometeorites in a single search.

When you have discovered a promising roof, send an email to the owner, explain the project, and politely ask for permission to take a scientific dust sample from the roof. Some owners will deny access for safety reasons. Others might be interested and welcome you for a brief sampling. Climbing a ladder and working on a high roof is a potential risk, so do not take any chances.

2 SEVEN STEPS TO HEAVEN

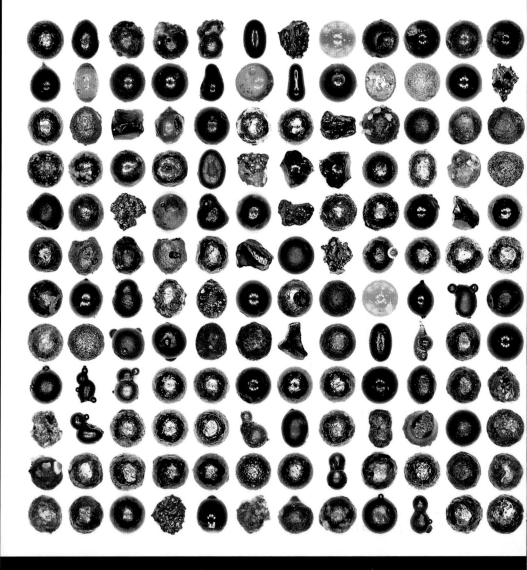

HOW DO WE FIND A MICROMETEORITE hidden among billions of other particles? Micrometeorites offer two clues that will help us in the process.

First, most micrometeorites contain small amounts of iron and nickel. These we can extract from the rest of the particulate we gather with our weapon of choice: a magnet. In the cosmic spherule collection from the South Pole, for example, approximately 80 percent of the micrometeorites are magnetic.

The remaining 20 percent of spherules are nonmagnetic. Here, the metal content has been evaporated completely by the frictional heat created during atmospheric flight. These spherules we have to search for in a different way, which we will get to later.

Meanwhile, the short road to finding 80 percent of micrometeorites is a magnet. Any type will do, but the stronger the better. I use a handheld neodymium magnet that measures 40 millimeters in diameter and has a hook in the center that serves as a grip. The magnet shown on page 20 is the one I have used to find all the urban micrometeorites in my collection.

The other property that will help us separate micrometeorites from terrestrial particles is their size. Most cosmic spherules are between 0.2 and 0.4 millimeter in diameter. Magnetic extraction plus screening for size narrows down the possible candidates in a dust sample sufficiently to start searching with a microscope for the proverbial needle in the haystack. An iron needle can be found relatively easily in a haystack if you have a magnet. But how do we use the magnet out in the field, in the rain gutter or on the roof?

Not only do we want to catch micrometeorites with a magnet, we want to release them again in a controlled way at the right moment, in the right place. And we want to avoid contamination of particles from one field search to the next. There are seven steps in the process of separating the micrometeorites from the terrestrial dust.

The spherule content from an average tablespoon of urban dust from any place on Earth. Numerous rounded objects, but which is which? And are there any extraterrestrial particles here?

A powerful neodymium magnet with a practical hook to hold it. This is the magnet with which the author has found the micrometeorites.

STEP 1. Put the magnet inside a small plastic ziplock bag; this is bag number 1 (see photo on page 21). Hold the magnet by the hook from the outside. The plastic bag keeps the magnet clean and avoids transmitting particles from one place to another. Change it after each field search, or more often. These small plastic bags are sold in packages of a hundred for a few dollars.

STEP 2. Put your entire hand with the magnet inside a larger plastic bag. The outside of bag number 2 will be the magnet's contact with the ground. Rough particles in the dust will act like sandpaper and make small holes in the plastic. If complete separation of particles from one place to another is required, bag number 2 should also be changed often. Bag number 2 must be held tightly around the magnet, so it gets as close contact with the dust on the ground as possible.

Step 1. Put the magnet inside a small plastic ziplock bag.

Step 2. Put your hand with the magnet and first bag inside a larger plastic bag.

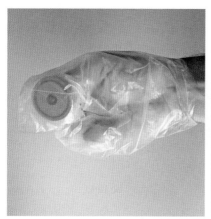 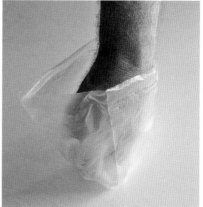

Step 2, continued. Make sure the plastic is snug around the magnet. This will be the magnet's contact with the ground.

   Let the magnet contact the sampling area on the ground through the two plastic bags. Magnetic particles will get stuck outside plastic bag number 2.

Step 3. Hold another small plastic ziplock bag in your free hand, ready to collect the magnetic particles that cling to the outside of the second outer bag.

STEP 3. Hold a small, empty ziplock bag in the other hand, ready to collect the dust sample. This is bag number 3. Put the magnet with the dust particles ("magnetic catch") from the ground at the opening of bag number 3. Hold back bag number 2 with one hand while pulling the hand with the magnet out of bag number 2. When the magnet is removed, any magnetic particles will be released and fall into bag 3. Repeat the sequence until the required sample is obtained.

Use the magnet to take dust samples in places where small particles have accumulated over time. Wind and rain sort particles by size and weight (density). If the particles are moist, the magnet will not work because the adhesion of water is stronger than the magnet. In these cases, use a spoon to obtain a sample and put it in a plastic bag marked with the place and date. Later on, dry and rinse the sample (see step 5); then you may use the magnet.

Sampling dust on a flat roof is more efficient when you use a dust broom to gather loose particles before using the magnet. Hard lumps of dry dust should

Step 4. Label your magnetic samples with a permanent marker.

be broken apart. Around a drain or in a pitched roof's gutter, a stiff dishwasher brush can be useful. Alternatively, collect the entire contents in a large plastic sack and later rinse the sample by flotation (see step 5, alternative method).

STEP 4. Mark the field sample in plastic bag number 3 with a permanent marker. Write the finding place, date, and any additional notes. This may seem insignificant, but sometimes there is not time for an immediate examination of the field sample. Then it is important to have marked it properly for later.

It's advisable to follow the weather forecast and try to choose a *dry* day for your field search. This will save extra work. If the sample is moist, the magnet will not extract the magnetic particles. If you must collect a moist sample, an alternative solution is to gather it in a plastic sack and clean it later, either by flotation or magnetic extraction followed by rinsing (see two methods described in step 5).

STEP 5. This step is all about cleaning the sample with water. On a dry, sunny day it is easy to extract magnetic particles in situ, whether the field search is conducted in a rain gutter or on a flat roof.

For the step 5 rinsing process, use a bowl, hot water, some dish soap, and a plastic spoon. The micrometeorites are mainly black, so whenever possible I use white porcelain or plastic buckets, bowls, plates, and plastic spoons for the entire process.

Fill the bowl halfway with the hot water and some dish soap. Pour in the sample and stir; as the water turns black, stir more. Any organic matter will float, while the mineral particles will sink. Stir again. After a while, let the mineral particles sink and slowly pour off the froth and dirty water. Make sure to keep the heavier particles at the bottom. Repeat this process until the organic particles are gone and the water remains clear.

Step 5. Now it's time to clean the sample. If one constituent of the particulate is bird droppings, an initial disinfecting in rubbing alcohol is recommended. Prepare for the cleaning by putting the dust sample into a bucket or bowl.

Pour hot water and some dish soap over the sample and stir gently with a wooden stick or a plastic spoon.

Allow the heavier particles to sink. Then gently pour off the froth and dirty water. Repeat the process until the water remains clear.

STEP 5 (FLOTATION ALTERNATIVE). Sampling for micrometeorites on a dry day is efficient. After a while with a magnet on a roof following steps 1 through 4, you might extract a sample of magnetic particles from the place where they accumulated. When the time comes to rinse the sample (see step 5 above), a couple of grams will take only a few minutes. However, if conditions on the roof are moist and the sample is a plastic bag full of mud, we must clean it in a different way: by *flotation.*

An old roof may have accumulated so much debris that the lumpy sediments in the corners or around the drain never dry out. Or you may only get permission to take a sample from a roof on a rainy day. Or maybe you live in an area with a lot of rain. And on a pitched roof the rain gutter may be so full of windblown soil, moss, leaves, pinecones, or sand that it is always wet. In all four cases it is still possible to separate and collect micrometeorites.

In the case of the pitched roof, we must empty the entire contents of the gutter. This should be done once a year anyway, and many homeowners will even thank you for doing it for them. Follow ladder safety procedures and climb up to the gutter, then hang a plastic bag on the ladder. Wearing rubber gloves, use one hand to get out most of the content from the gutter while gripping the ladder with the other. Put the lumpy gutter sample into the bag and use a stiff brush to gather the rest of the mud from the bottom of the gutter. This may be

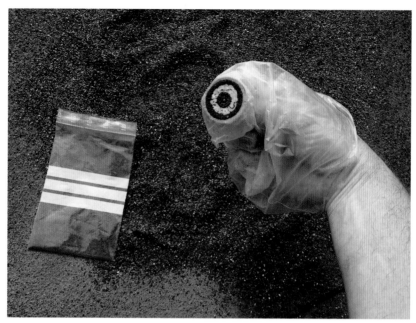

Sampling on a flat shingle roof. Despite the large amount of loose debris on the roof, the magnetic content is low, which makes the subsequent rinse and screening for size relatively easy.

where the heavier micrometeorites are hiding. It is a dirty job, but someone has to do it, and the reward is celestial.

Move the ladder to the next part of the gutter and continue until you are satisfied with the sample. When you are finished, mark the bags with the location and the date using a permanent marker.

On a flat roof you can use a coarse broom and a shovel. Be careful not to puncture the roof covering. A large roof may contain several hundred pounds of debris, so choose your battles wisely. Remember the signal-to-noise ratio: a lot of mud does not necessarily mean a lot of micrometeorites.

Now it is time to process the sample. Find a place where you can work undisturbed with access to water, preferably a garden hose. This process is called flotation, the alternative version of step 5, but in fact it is just as much "sinkation": the mineral particles sink, and the lighter organic matter floats. I use plastic buckets and wear long rubber gloves.

Fill the bucket halfway with the gutter content and pour water over it until the bucket is almost full. Gently massage any clumps until they disintegrate to

Alternative flotation method: Put the contents from the rain gutter in a bucket (fill no more than halfway) and add some dish soap and lots of water.

Loosen the lumps with your hand, and let the heavier mineral particles sink, while the lighter organic matter floats.

Let the bucket rest for a while, then pour out the dirty water and floating organic parts before refilling with clean water.

Continue until all the organic matter has been removed and the water remains clear.

The sandy grains on the plate are cleansed. This sample consists of both magnetic and nonmagnetic particles of all sizes.

create a soft, muddy soup in the bucket. Gradually the organic content will float and the heavier mineral particles will sink.

Pour out the top layer of organic matter and the muddy water, making sure the bottom layer with the minerals remains in the bucket. The micrometeorites are small and need a minute without turbulence to sink. Refill the bucket and repeat the process until the water is clear and all the organic material is removed. This can require ten to twelve rounds of changing the water.

The sandy remains at the bottom of the bucket can be poured out to dry on a clean plate. Make sure you get even the smallest grains—micrometeorites will be among the finest particles. Do not despair if the mineral sample from several buckets of rain-gutter slush results in just a teaspoon of mineral particles (or less). It may still contain a lot of micrometeorites.

The sandy grains on the plate are cleansed. This sample consists of both magnetic and nonmagnetic particles of all sizes.

At times there are no visible organic particles in a sample, but all the same, the water may get black from the dirt particles measuring less than 50 microns (μm). Susan Taylor, who found the micrometeorites in the South Pole Water Well (SPWW), advises disregarding all particles smaller than 50 microns. The main purpose of step 5 is to rinse the mineral particles so that later we will be able to recognize the surface textures on each individual particle under the microscope. A clean sample is fun to search through; a dirty one is tiresome for the eyes.

STEP 6. Put the rinsed mineral sample on a plate to dry, making sure to get all the smallest particles from the bottom of the bowl. The total weight of the sample may have been reduced by up to 80 percent since rinsing, for example from 25 grams to 5. This is quite substantial, considering the total dust content on the roof before the magnetic extraction may have been 250 pounds (113 kg) in the first place.

The two properties of micrometeorites, which we use to separate them from terrestrial particles, are magnetism and size. When a rinsed magnetic sample is completely dry, we can proceed to step seven: *fractioning*, or screening, for size. This is also where we continue with the mixed mineral particles extracted from the wet rain gutter by flotation.

Step 6. Put the rinsed mineral sample on a plate to dry.

Step 7. Screening for size. Two strainers with different fraction: the strainer on the left has 1.5mm mesh, and the one on the right is 0.4mm. Most micrometeorites are smaller than 0.4mm and so are mainly in the smallest fraction, which fall through the finest mesh.

STEP 7. This is where we take advantage of the fact that there are more micro-meteorites between 200 and 400 microns than there are smaller or larger ones. The average cosmic spherule has a diameter of around 300 microns. To isolate the fraction of particles that fall into that size range from our sample, we use a sieve. An ordinary tea strainer has a mesh fraction of 500 microns, or 0.5 millimeter—micrometeorites would be among the fine particles to fall through such a strainer. Fold a piece of white paper in half and use it as a funnel to get all the particles from the sample into the strainer. Keep a white plate under the strainer while shaking it gently. The particles on the plate will now be in the 50- to 500-micron fraction. With some practice on step 5, the rinsing, this may be reduced even further, to 100 to 500 microns.

Now we are down to less than half of the weight of the sample from step 6— around just 2 grams in all.

Larger and finer sieves are available. Hardware stores often have a type that is 20 centimeters in diameter with holes of 400 microns, used in the kitchen for

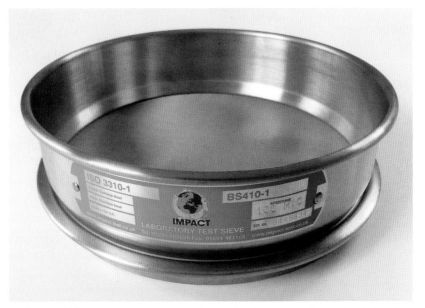

Professional laboratory sieves come in ultraprecise fractions and are splendid tools for the semiprofessional stardust hunter.

flour. With this type it becomes possible to reduce the dust sample even more, to a 100- to 400-micron fraction, and minimize the time spent on the microscopy. The fewer particles to examine, the less tiring for the eyes.

Professional laboratory sieves have ultraprecise fractioning, but these are expensive. Nevertheless, if you really get hooked on micrometeorite hunting, it is worth it. By using laboratory sieves in the 400- and 200-micron fractions, you quickly go directly to the main size for the micrometeorites. Still, keep in mind that there *are* micrometeorites in the sub-200-micron fraction and a few larger than 400 and even 500 microns.

When I invested in laboratory sieves in fractions of 150, 200, 300, 400, and 500 microns, I found that the number of micrometeorites I retrieved immediately increased. One reason is that under the microscope, depth of field is limited, and the more homogenous the particle sizes are, the easier and faster it is to examine the sample. If the sample consists of grains of various sizes, it is necessary to constantly readjust your focus, which takes more time and causes

A dust sample from a roof screened for size by the two kitchen strainers. On the top is a plate with the fraction larger than 1.5mm. The plate in the middle is between 0.4 and 1.5mm. The bottom plate holds particles smaller than 0.4mm—where most micrometeorites will be found.

you to miss details. Still, a tea strainer will do when starting out. I found my first 900 micrometeorites with one.

It must also be mentioned that there are other ways to rinse and fraction. Some prefer to do the entire process under wet conditions, including the subsequent examinations under the microscope. The advantages are that you do not have to wait for the sample to dry and you can do the entire process in one sequence. One disadvantage is you need a laboratory or at least a designated room or area.

Others prefer to use a sonicator for rinsing. A sonicator uses sound waves to agitate and separate particles in a sample and is an excellent tool that might expand the hunt for micrometeorites in the future.

Now we are almost ready for the exciting microscopy, but first we put the sample on a small plate. Again, we apply the magnet, and, voilà, this time only half of the sample is caught by the magnet—the rest remains on the plate. How is this possible? Have the particles lost their magnetism?

The explanation is simple: the original dirty sample from the roof contained magnetic particles held to nonmagnetic particles by dirt. During the rinsing they were separated.

This magnetic sample is now down to 1 gram, or approximately 0.001 percent of the initial mass on the roof. The methodical process has filtered away 99.999 percent of the particles. Now we are ready to search for the needle in the haystack.

3

YOUR PHYSICAL COLLECTION

MICROSCOPY AND HANDLING

With the magnetic dust sample rinsed and screened for size, it is time to put the particles on a small petri-type dish under the microscope.

To find the micrometeorites among the other magnetic particles, we must first get to know them so we can recognize them in the sample, like a familiar face in the crowd. This is explained in part 4, followed by a presentation of the other types of spherules you will encounter in the urban dust. In short, though, look for the aerodynamic forms and the characteristic surface textures, barred or porphyritic olivine, cryptocrystalline "turtlebacks," and so on.

I recommend a stereomicroscope because the three-dimensional image it creates is the best for recognizing and identifying the fine surface textures of the particles. There are low-budget stereomicroscopes for sale online, but professional microscopes are better, albeit more expensive. I bought a used Zeiss binocular microscope 25 years ago for $400 and it will last a lifetime.

The microscope's most important feature is its magnification. To examine larger particles, around 1 millimeter, 25× magnification is fine. But micrometeorites are smaller; I use 63× magnification to see the surface textures on individual particles. Good USB microscopes are available on the Internet offering magnification up to 200×. These microscopes reveal a lot of morphological details of small stones, but as magnification increases, depth of field decreases proportionally and requires stronger light than that needed for observation at lower magnifications.

Furthermore, the lack of stereoscopic view makes it difficult to use a USB microscope for only that. It is, however, splendid for taking color photographs directly to a hard drive. If photographing in magnifications larger than 200×, it is necessary to take a sequence of photos while changing the focus area between each exposure, then "stack" the photos with software.

Rings of dark stardust around the Andromeda Galaxy. We are in the second generation of star formation, and the cosmic dust is getting heavier. In micrometeorites we can find platinum group elements directly from a previous generation of supernova explosions. *Jan Inge Berentsen Anvik*

A binocular microscope gives a three-dimensional view of particles, which makes it possible to recognize the characteristic surface textures of micrometeorites.

Contrary to common belief, micrometeorites are not metallic spheres. They are small *stones*. They are often a bit darker than the rest of the particles in a sample. Because they have fallen onto a rooftop where there's little human activity, these small particles are *complete* objects; whereas most terrestrial mineral particles are cracked, broken, and worn, micrometeorites are undamaged aerodynamic stones. Of all the micrometeorites I have found, only a single stone was split in half. The rest were pristine.

A 1-gram magnetic dust sample in the 200- to 400-micron fraction will contain hundreds of thousands of particles—examining each one is time consuming. I recommend putting only 0.5 gram, perhaps even less, on the dish at a time. Spread the particles thin, start at one end, and work your way through the sample.

Beginners especially will find many interesting things to see—colorful crystals, metal spherules, glass and organic surprises—and the examination will take a long time. Try to enjoy it. After all, you are exploring something new. In this part of the search for stardust, patience is of the essence. Relax and turn on

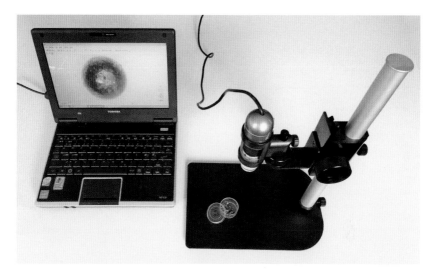

A low-budget USB microscope can take color photos and save them directly to your computer's hard drive. With these photos of your findings you can get micrometeorite verification from more experienced collectors and gradually build your own photo database of micrometeorites (and micrometeowrongs).

One micrometeorite magnified among non-micrometeorites. *Scott Peterson*

An average sample from a roof, cleaned and screened for size. Can you spot any spherules?

While the author prepared breakfast on a porch one sunny morning in 2009, he noticed a small black dot suddenly appear on the table. The unexpected discovery was the starting point of his search for stardust. Seven years later he found his next micrometeorite and went on to become the first person to find micrometeorites in populated areas. *Ryan Thompson*

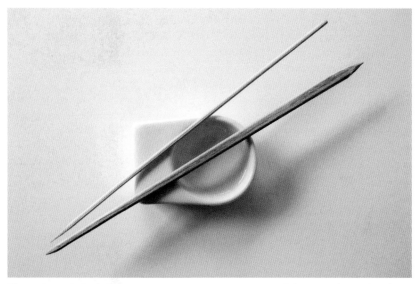

The main tools for handling micrometeorites are water and sharpened wood sticks—a dry stick to move the particles around under the microscope and a moist one to pick out interesting finds.

music, a podcast, or the radio. With time you will gain experience and recognize more of the usual suspects.

Like a telescope, a microscope gives us a big picture. With experience the search may become almost subconscious, and it will become possible to scan tens or perhaps hundreds of particles at a time without even focusing on each individual object as you move the sample slowly across the field of view. Continue until you have checked the entire sample, then scan quickly through it one more time.

When necessary during microscopy, I use a sharpened wood chopstick to move individual particles around so they can be inspected from all sides. Wood is excellent for avoiding static electricity, which is a problem when using needles of metal, glass, or plastic. Furthermore, metal needles leave ugly marks on the beautiful micrometeorites.

I also have an even sharper bamboo stick that I use to pick out interesting objects. I dip the tip in water, and when it is moist, the water adhesion holds the particle onto the stick just long enough to move it to a new place for archiving or photography. It cannot be too wet, or the tiny stone will not let go when it is at its new destination.

When searching for micrometeorites under the microscope, or handling microscopic dust particles, always have a white sheet of paper on the table under the microscope or under the USB microscope (camera) and the small archive boxes (see pages 43 to 45). Occasionally a particle will fall off and get lost, but with strong light on the desk, it might be possible to find it on the white paper.

Alternatively, use the magnet and see if you can catch it once again. With the naked eye a micrometeorite looks like a tiny black dot barely perceptible by human vision. One time I lost a nice micrometeorite while moving it to photograph it. I put on a lot of effort into finding it again and searched all over with the magnet. When I examined what the magnet had picked up from the floor, I found no fewer than three micrometeorites!

CURATING YOUR COLLECTION

Searching for stardust is an exciting and interesting hobby. At the same time, it is cutting-edge science. The possibility for new discoveries is very real. Despite a lack of a commercial market for micrometeorites, their scientific value is sky high. However, there are still more questions than answers about cosmic dust particles. Where do they come from? What are their parent bodies?

When unmelted micrometeorites enter the atmosphere, they contain complex organic molecules—amino acids and water—that are the building blocks of life. What role, if any, did micrometeorites play when life arose on Earth? And is the omnipresent cosmic dust playing the same role in other parts of the universe, on the many recently discovered exoplanets, for instance? Nobody knows where Earth's water came from. Did it arrive under the radar with the micrometeorites?

Big questions are connected to these tiny particles of cosmic dust. We must therefore carefully curate our micrometeorite collections to maintain their scientific potential. This can be done with a few simple steps that will increase the value of your collection dramatically.

First, start a handwritten journal, even before you've found cosmic spherules. Use a hardcover notebook or make one yourself from a ring binder.

Give each field search a unique number and let this number follow all further documentation of the findings from that search. This step will link micrometeorites to where they were found and when. Allow a half page for each field search. Make it simple. Start with the field-search number, date and place of the search, and whether the search was done with or without a magnet.

If you like, continue with whatever additional information you have. For instance, the size of the roof, the type of decking, the age of the roof, or how

With a digital scale you can measure the rate of micrometeorites per gram in different samples and assemble data regarding which field searches are the most successful. You can also compete with yourself and other collectors. The author's record is 27 micrometeorites per gram.

long since it was cleaned. If you experiment with new methods of sampling or rinsing, make sure to note them in the journal. If you weigh the samples, write the initial gross weights and the weights after rinsing and fractioning. With this information you can compete with yourself and others for the highest number of micrometeorites per gram of sample.

Record the information about what you find in the sample *while* doing the microscopy. Sketch interesting findings and new discoveries in the journal. This is not an art contest but an efficient way to remember. When you start to find the micrometeorites, make a quick classification of each (barred olivine, cryptocrystalline, etc.) and mark in the journal how many of each type you find, and preferably also in which fraction.

Keep track of your micrometeorites by giving them individual numbers. One way to do this is to use a combination of the unique field-search number and a hyphen plus the number of the micrometeorite in that specific sample. If this is field search number one hundred, for example, and you find two microme- teorites, they can be numbered 100-1 and 100-2. Put a letter prefix in front of the number, for instance your initials plus "MM" for micrometeorite, to distin- guish your collection from others'. Thus, Jane Doe's first micrometeorite on her thirty-fifth field search is JDMM 35-1.

Others use a numeric system based on how many micrometeorites they have found in all, starting at 1, in combination with a letter prefix and the name of the finding place (instead of the field-search number) after "MM." These individual numbers make it possible to refer to micrometeorites and exchange samples and findings without losing track of an object. As micrometeorite collectors begin to trade, it is imperative to know the provenance of every tiny space rock.

On your computer you should keep a separate folder for the documentation from each field search, named by the individual field-search number plus the name of the place. In each folder you can put all photos related to that field search: portraits of findings and snapshots from the roof during the field search, including the sample in situ. As you advance in your searches, include things such as SEM images and chemical spectra from energy-dispersive X-ray spectros- copy (EDS) analysis. This enables us to examine all documentation from each field search efficiently and compare it with others. Without a photo database organized like this, I would never have succeeded in identifying all the different terrestrial contaminants I have, which in turn made it possible to break the code of micrometeorites.

I recommend sharing information about your findings, images, stories, and analyses with the growing community of stardust hunters around the world. There are enough micrometeorites for all of us, and every day 100 new tons fall to Earth. If we manage to retrieve only a small fraction of this, the micrometeor- ite research will explode, and we might get closer to the larger mysteries of the universe. Where do we come from? Who are we? What is our future?

If you really want to excel in this field, consider keeping a spreadsheet in which all your data is searchable: the history of each micrometeorite, classifica- tion, photos, SEM images, chemical analysis, and so on.

Storing or archiving your micrometeorite collection has advantages over other types of collections. After all, micrometeorites are small and do not take up much space. This is why rock hounds envy us. You should be able to find

A tray with twelve field-search boxes. Note the specific number on each box. At times the findings from several locations can be placed in the same box, but most often complete separation is desired.

every single micrometeorite in your collection with ease and store them in very little space. One way to archive it is to store all findings from one field search in a small plastic box with tight lid and the field search number written on it (see photo above). You could take it a step further and affix a white sticky note to the bottom of the box and organize the micrometeorites in rows. Others use small laboratory test tubes for archiving. In any case, let the field search number follow the findings, as well as each micrometeorite's individual registration number.

A good way to store a micrometeorite collection is with this type of slide made for microfossils. By gluing a small sticky note in the recess, it is possible to organize the micrometeorites in lines. A number at the back of the slide references a spreadsheet or database with full information about the micrometeorites on this slide. It should be possible to find any given object in the collection easily.

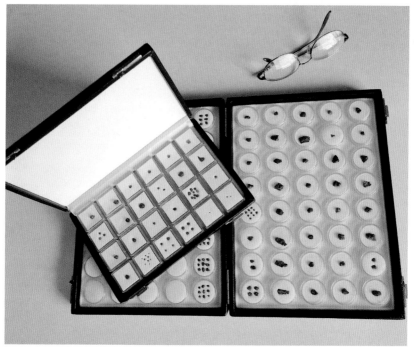

Showcases made for gems work well for micrometeorites.

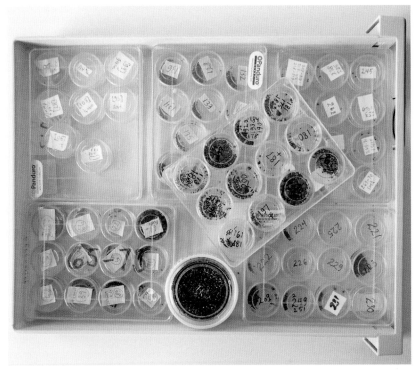

A small chest of drawers with room for five trays of twelve boxes each. One drawer can hold the findings from sixty field searches. The point is to organize your collection so you can quickly find one particular micrometeorite and keep track of its history.

Microfossil collectors use a special type of slide for archiving and displaying their collections. These slides are made of cardboard, have a shallow recess for the microscopic fossils, and are covered with glass held in a metal frame. These slides also work for micrometeorite collectors (see photo on page 44). They take up very little space and are durable and easy to store in a reliable system. As with small boxes, it is also possible to attach a small white sticky note or label to the bottom of the hole and arrange the micrometeorites in rows. Write the individual micrometeorite number on the back of the slide and enter the numbers in a searchable database. This makes it possible to keep track of each micrometeorite's provenance and find a specific micrometeorite in your collection within seconds.

All the information you record in your journal and database will add up to very interesting data in a field in which there is still very little available information. How much cosmic material is really reaching the ground over time? What kind and in what fractions? Where does it accumulate, and where do we find the most? Practically everything we know about cosmic dust particles is based on a teaspoon of dust from Antarctica, which means the data points are few. This is changing. There will be new discoveries in this field, and they may arise from *your* journal.

EXPERIMENTAL FIELDWORK

In just a few years the results of micrometeorite fieldwork have improved tremendously. This despite the fact methodology has gone basically unchanged, with only a few minor modifications, since the verification of the first urban micrometeorite.

So why the improvement? The main difference is experience, which makes it a lot more efficient. For example, today I spend more time in advance searches for the right roofs and less time on microscopy. We are still in the infancy of the micrometeoritics and should continue to develop the methodology. I strongly recommend experimenting with new hunting grounds, new ways of using magnets, and using our imaginations to develop new tools.

At times we tend to think that everything has already been invented and discovered. Of course, this is not true. A new discovery might be right in front of us, barely camouflaged by our habitual thinking. To continue to improve the results in the hunt for micrometeorites, we must try new methods and expand the hunt.

It is possible to monitor one specific area thoroughly; for example, a particularly favorable vinyl roof with relatively high security walls or the rain gutter of an especially promising pitched roof. Wash and clean it, then examine it for new particles at certain intervals, for instance once a month. It's hard work, but it will result in new data about cosmic dust.

Also, experiments can be partly automated. For instance, maybe a robotic lawnmower equipped with magnets instead of spinning blades underneath can graze quietly for extraterrestrial particles.

A metal detector can also be modified into an electromagnet. It is better for your back to search for micrometeorites while standing up than bent to the ground. There are already "magnetic brooms" for sale, which may be useful for quickly searching a flat area for freshly fallen particles.

The larger and older a roof, the better. The darker gray roofs are covered with shingle; these are also possible hunting grounds for micrometeorites, but magnetite content in slate at times makes the examination of the sample more time consuming. The PVC-covered roofs are the most efficient.

A vacuum cleaner can be used to efficiently suck up a lot of dust from a flat surface or a dry rain gutter. The magnetic particles in the vacuum's bag can be extracted in the same way as any other sample. Or the vacuum cleaner can be modified with magnets and meshes to catch magnetic particles in a certain fraction.

There are many ways a rain gutter can be modified to serve as a micrometeorite collector. Place strong magnets at low points and inspect them at regular intervals. Modify the downspout with a removable section equipped with a rack of meshes in specific fractions, catching the particles washed down by the water. Or place a bucket under the downspout and let it collect particles over a period.

To improve the flotation or rinsing process, construct a magnetic safety net or magnetic trap at the edge of the bucket or bowl while pouring out the dirty water, or modify a funnel with strong magnets. This way you might save more of the extraterrestrial particles. (I am not sure, but I suspect some scoriaceous micrometeorites may float.)

Extremely strong magnets are available for sale. A couple of these attached under a baby stroller, for example, should catch freshly fallen micrometeorites from the ground on all sorts of flat surfaces, even dry grass. Or wrap one in

On the trail of stardust in the Sahara. In a remote place like this, with hardly any human activity, one would think that it would be easy to find micrometeorites. But it was not—most sand grains are dark, rounded, and covered in a "desert varnish," making it impossible to recognize the surface textures. *Morten Bilet*

waterproof plastic and search underwater, where the signal-to-noise ratio will be very favorable. It was on the ocean floor that the very first micrometeorites were discovered, by John Murray during the oceanographic *Challenger* expedition in the 1870s.

As many glaciers around the world melt, cosmic particles that collected in the ice for thousands of years should be revealed in the melting zones. By collecting and melting dirty ice in these areas and filtering the water, it should be possible to extract micrometeorites—and, likewise, by filtering the seasonal runoff from a glacier in the summer. This experiment has been successful in Antarctica. Despite the weaker signal-to-noise ratio found in glaciers outside Antarctica, it will still be better than in the dust on an urban roof.

You can also construct micrometeorite traps with tarps. Set up a tarp such that the corners and edges are held off the ground and make a small drain hole in the middle, held down by weights. Place a bucket under the hole. Perhaps also make a drain in the upper part of the bucket to lead the water away in a controlled manner.

If you plan on building a new house and are using an architect, ask whether the roof can be optimized for micrometeorite collecting. This can be combined with solar panels.

There are professionals who maintain the decking, air conditioning, antennas, and other devices found on roofs, as there are professional rain-gutter and roof cleaners, all of whom have access to the best micrometeorite collection sites in the world. Cooperation with these people may be fruitful.

To search for nonmagnetic (glass) micrometeorites, it is possible to construct a vibrating pitched table that collects the round glass spherules before they roll down and off. Such tools are available for sale, but they are expensive.

Mechanical street sweepers collect thousands of tons of street debris every year, in which there are bound to be large amounts of micrometeorites. In some countries this debris is discarded because of possible storm-water pollutants, but it should be possible to rinse, fraction, and recycle the various particles—and at the same time retrieve a lot of micrometeorites.

The ultimate micrometeorite processing facility, however, would be an industrial line—an upscaling of the steps used in the rinsing and flotation processes. The raw material, for instance tons of debris from roofs, would go in at one end and proceed through flotation in large pools, then rinsing, drying, magnetic extraction, and screening for size. Perhaps a vibrating pitched table and even artificial intelligence (a modified facial-recognition system) would pick out promising extraterrestrial candidates. The micrometeorites would roll out the other end and the rest of the material would be composted.

The time for the exploration of the micrometeorites is now. Whatever method you use, enjoy the irony that it is possible to find the most exotic particles in our solar system, and the oldest existing matter, on the nearest roof.

4

WHAT WE FIND IN THE DUST

PREVIOUS MICROMETEORITE STUDIES have concentrated primarily on the various aspects of chemistry and classification. For this, backscatter SEM section imaging has been the most common way of depicting micrometeorites. The chondritic spectrum is the chemical composition of the *entire* solar system, which is found again in the cosmic dust grains, down to the minor trace elements. How is this possible? The explanation is simple: the cosmic dust was here first. The solar system 4.56 billion years ago was made of a dust cloud, the *Solar Nebula*. In a micrometeorite, the entire solar system is mirrored chemically in one microscopic particle.

To identify micrometeorites in populated areas, however, it is necessary to know what they *really* look like—in other words, their *morphology*.

The following sections present a photographic overview of the most common types of spherules we encounter in dust samples. We will begin with micrometeorites, because only when you know what micrometeorites look like is it possible to recognize and find them.

Next are the two rare types of extraterrestrial spherules that are *not* micrometeorites: *ablation* spherules and the enigmatic *chondrules*.

Finally, a longer section presents all the most common types of *terrestrial* spherules, both manmade and naturally occurring.

Compare your findings with these images. There are additional photos in the album section of Project Stardust's Facebook pages and in this book's predecessor, *In Search of Stardust*.

The purpose of this chapter is to give you an idea of the various aerodynamic forms and characteristic textures of cosmic stones so you can find your own.

Chemically, the various types of cosmic spherules are remarkably homogenous (see page 130). The differences among them are caused mainly by the peak temperature and quenching profile during their atmospheric deceleration. The illustration on page 52 shows the effect of temperature on cosmic dust particles, from unmelted to glass.

A collage of nine micrometeorites on blue background. *Jan Braly Kihle/Jon Larsen*

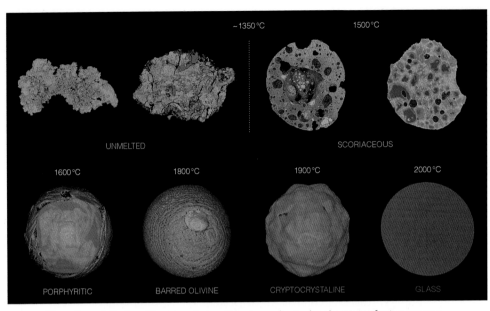

~1350°C   1500°C

UNMELTED   SCORIACEOUS

1600°C   1800°C   1900°C   2000°C

PORPHYRITIC   BARRED OLIVINE   CRYPTOCRYSTALINE   GLASS

The effect of frictional heating during the atmospheric deceleration of micrometeorites, from unmelted (blue) to completely melted (red). These are the various types in today's classification of micrometeorites.

The micrometeorites on the following pages are marked with catalog numbers from my collection. They have the prefix NMM (Norwegian Micro Meteorite) in front of the number. A scale bar has not been added to the picture because the variation in size is negligible; all are around 0.3 millimeter. Also, the classification of the micrometeorites is not absolute, as there are transitional forms among all the various types; occasionally, for example, one micrometeorite will display features of three different types.

All the micrometeorites shown here were found using the methods described in this book, and they were the first urban micrometeorites found. Furthermore, the photos on the following pages are the first high-resolution color photographs taken of micrometeorites, created with the instrument Jan Braly Kihle and I developed especially for Project Stardust (and described in part 5). Enjoy!

THE MICROMETEORITES

BARRED OLIVINE

The most common type of micrometeorite and statistically the first one you are most likely to encounter in your dust sampling is the barred olivine (BO) type. These particles contain the mineral olivine and small amounts of magnetite inside the glass. As with most micrometeorites, the olivine is of the magnesium variety (forsterite).

BO spherules were fully melted, but upon cooling, magnetite and olivine crystals formed. The olivine crystals are aligned parallel to one other over large fractions of the sphere. These crystal grains are quite large in comparison to the overall size of the micrometeorite, so BO spherules are called *coarse-grained* micrometeorites.

The surface has recognizable striations, at times with metallic magnetite "Christmas tree" crystals sprinkled over it. These are formed by chemically bound iron from the micrometeoroid reacting with oxygen in the atmosphere. During microscopy, the parallel-oriented crystal plates reflect the light in a characteristic way, useful for identifying the type.

Between 5 and 10 percent of BO micrometeorites have a metal bead on their surface. At times it is possible to see how the metal has served as the nucleus for the recrystallization of the micrometeoroid in a dynamic interaction with the aerodynamic forces.

Barred olivine (BO). This is the most common type of micrometeorite. Note the aerodynamic forms and characteristic striations on the surface textures. Some have visible magnetite "Christmas tree" crystals on the surface.

NMM 705

NMM 930

NMM 794

NMM 867

CRYPTOCRYSTALLINE

Cryptocrystalline (CC) micrometeorites are statistically the second-most common type you will find in a field sample. They are mainly glassy particles with fine-grained crystallites too small to recognize as individual grains. They may have fully developed oriented and aerodynamic forms at times elongated with a metal bead in the front—or, if the particle had spin during its atmospheric deceleration, metal beads in both ends or multiple beads around the rock in places consistent with spin and the centrifugal force that were at work during atmospheric flight. If a metal bead has fallen out, it may have left a hole in the stone. One variety of CC micrometeorite is the characteristic "turtleback," round stones with humps evenly distributed around the spherule, not unlike a turtle shell.

Fine-grained cryptocrystalline (CC). Some of these have metal beads, some are the so-called "turtlebacks," and some are translucent. Note, however, the characteristic aerodynamic forms.

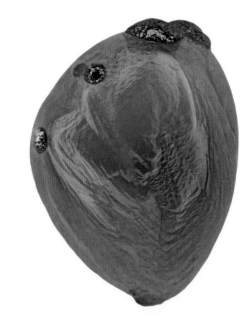

NMM 1075

NMM 1196

NMM 362

NMM 244

NMM 1359

NMM 626

NMM 628

NMM 646

NMM 716

NMM 650

NMM 752

NMM 874

NMM 878

NMM 928

PORPHYRITIC OLIVINE

Porphyritic olivine (PO) micrometeorites are large olivine (forsterite) crystals in glass that suffered a relatively low peak temperature during their atmospheric deceleration. They display morphological varieties from a mosaic of evenly distributed small crystals of cumulatively increasing sizes to just a few extremely large crystals or even possibly a single olivine crystal. They range in color from black (most common) to brown, green, and colorless. At times they have an irregular surface that may cause confusion in their identification, and they may be mistaken for asphalt spherules (see page 93). PO micrometeorites are often vesicular with visible metal beads on the surface, squeezed in between the olivine crystals.

Porphyritic (PO) olivine. Note the large crystals of the mineral forsterite. Some of the POMMs have nickel or iron beads, and some are translucent, greenish or brownish.

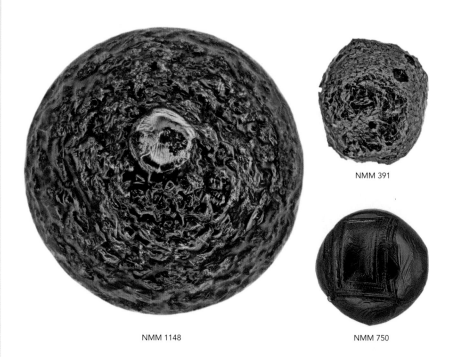

NMM 391

NMM 1148

NMM 750

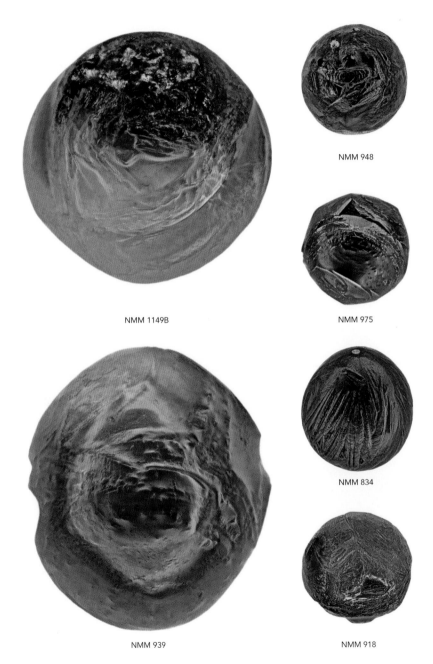

NMM 948

NMM 1149B

NMM 975

NMM 834

NMM 939

NMM 918

MICROMETEORITES / GLASS

GLASS

Glass, or V-type (vitreous), micrometeorites are the third-most common type but also difficult to find due to their lack of magnetism. They have suffered the highest peak temperature of all the most common types of cosmic spherules, up to 2,000°C (3,600°F), and all or most of their iron and nickel has evaporated, resulting in a sphere of amorphous glass. Colors range from colorless to brownish and green tints and even a rare bluish color.

Glass micrometeorites may be highly vesicular, even containing the remains of a burst glass bubble. Occasionally there is a metal bead on the spherule that is the remains of the former differentiated metal core pushed forward to the front surface by inertia during atmospheric deceleration. If there is a metal bead, we may catch the glass micrometeorite with a magnet. On some glass micrometeorites a beginning recrystallization around the metal remnants served as nucleus for the crystallization.

Glass (V type, vitreous). These micrometeorites suffered a very high peak temperature during atmospheric flight, and most of their metal content evaporated. Each of these six examples, however, has the remains of one or more metallic cores at its surface, pushed forward by inertia during the atmospheric deceleration but kept from escaping the spherule thanks to surface tension. Without these metal remains. these micrometeorites would not have been caught by the magnet. Approximately 20% of cosmic spherules are nonmagnetic glass micrometeorites.

NMM 789

NMM 1171

NMM 836B

NMM 739

NMM 1116

SCORIACEOUS

When a cosmic dust particle does not reach a peak temperature higher than around 1,350°C (2,500°F) during atmospheric entry and subsequent deceleration, the particle barely melts. Volatile elements expand and start to escape in the form of gas bubbles, resulting in a highly vesicular, or scoriaceous, micrometeorite (ScMM). These are dominated by micron-sized spherical olivine crystals, and transitional forms to porphyritic olivine have been observed.

With existing methods of extracting micrometeorites, it is not easy to find these. This is due partly to their lack of aerodynamic forms, which makes them difficult to recognize, but it is possibly also because they, like volcanic scoriae, have a low density and may even float. With improved methodology, and perhaps with the use of magnets during rinsing, we hope to find more of these interesting stones.

Scoriaceous (Sc). These micrometeorites from the NMM (Norwegian Micro Meteorite) collection are barely melted and contain unmelted relic grains from the original micrometeoroid.

NMM 1155

NMM 828

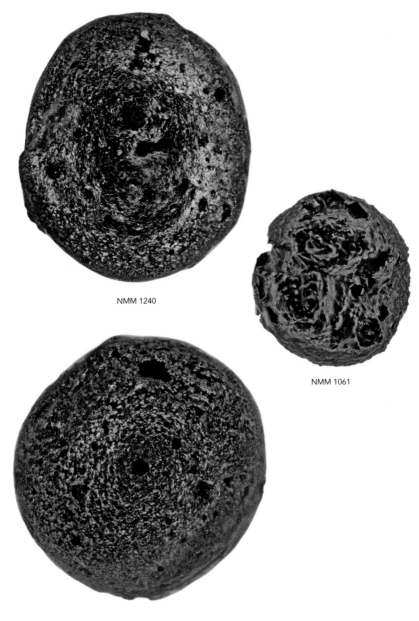

NMM 1240

NMM 1061

NMM 1047S

G-TYPE

G-type micrometeorites are glass with interstitial magnetite. The magnetite crystals may be euhedral (with well-defined faces) or cruciform (X-shaped). These micrometeorites are rare—or it is at least a bit tricky to identify them as extraterrestrial due to their complete lack of aerodynamism. Literature about micrometeorites contains little info about the G type. Scientists have not determined how they are formed. Having found a few myself and studied them under an electron microscope, I can say this is an enigmatic type of which I hope to find more to study more closely.

OTHER TYPES

There are other types of micrometeorites, both cosmic spherules and unmelted porous stone structures. The best known, still not identified in urban dust, is the **I-type** (iron oxide) micrometeorite. We are still searching for a method to distinguish these from I-type industrial spherules (see page 74), which look similar. Judging by samples taken from the South Pole Water Well, approximately 2 percent of micrometeorites are I-type cosmic spherules.

CAT spherules are enriched in calcium, aluminum, and titanium and have been found in the Antarctic, but they have not been reported anywhere for a while. Unlike most other micrometeorites, they are white. Described examples are either cryptocrystalline or barred olivine. One hypothesis is that they originate from a lunar incident, but this is speculation.

Some micrometeorites get a soft deceleration after the atmospheric entry and survive to the ground practically unaltered. These are called **unmelted** micrometeorites, and due to their lack of aerodynamic forms they are difficult to distinguish from terrestrial mineral grains. I have still not managed to find one. They have been found in Antarctica and may show some interesting phenomena. According to French micrometeorite pioneer Michel Maurette, unmelted micrometeorites display signs of having contained 12 to 15 percent water at atmospheric entry. And even more surprisingly, they also contain complex organic (nonbiogenic) molecules, including amino acids. A potent cocktail for life.

The so-called ultracarbonaceous Antarctic micrometeorites, or **UCAMMs**, contain up to 75 to 90 percent organic components. Embedded in this carbonaceous matter are small and complex assemblages of fine mineral grains and glass that resemble glass with embedded metal and sulfides (GEMS) that were first found in the interplanetary dust particles (IDP).

A fraction of micrometeoritic material consists of interstellar matter, called **presolar grains**. This is the oldest matter in the universe. It is estimated that 0.1

percent of micrometeorites are interstellar, but that number might be proven too low. One tenth of a percent does not sound like much, but it adds up to 90 kilograms (200 pounds) of interstellar matter falling to Earth every day. Some of this may be found on the nearest roof.

An incident in October 2017 made it clear that interstellar matter is not so rare on Earth as previously assumed. Without any warning, a 400-meter-long (130 feet) cigar-shaped object plunged at a 90-degree angle toward the ecliptic plane of our solar system's planets from somewhere near the blue star Vega. Its speed was a dazzling 60,000 miles (95,000 kilometers) per hour, the fastest interstellar object ever observed. The elongated object flew toward the sun, took a bow, and accelerated out into deep space in a completely different direction than it had come from. One scientist stated that "Unless the object suddenly slows down, there is no reason to assume it is a spaceship." Observations from the Very Large Telescope (VLT) in Chile revealed that the object had "a high content of metal." This was the first time an interstellar object like this was observed from the ground, but it is estimated that a large interstellar object visits our solar system as often as once a year. On the opposite end of the scale, we may enjoy small crumbs from outer space in the form of micrometeorites.

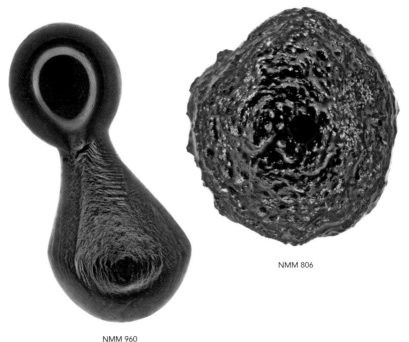

NMM 806

NMM 960

NMM 799

Other types of micrometeorites. NMM 799 (above) is a fine-grained chondritic spherule with large interstitial chromite crystals (the triangular skeletal octahedrons or tetrahedrons) plus magnetite "roses" on the surface. NMM 806 (opposite, right) is a G type, with magnetite crystals in glass; note the lack of aerodynamic properties. NMM 960 (opposite, left) is a rare cryptocrystalline droplet with a translucent brown glass sphere where the tail used to be.

EXTRATERRESTRIAL, BUT NOT MICROMETEORITES

ABLATION SPHERULES

It has been calculated that an average meteoroid loses around 85 percent of its mass during atmospheric flight. This erosion is called *ablation*. The term *ablation spherule*, meaning a melted spheroid micro object rubbed off a meteoroid by atmospheric friction, was previously often used to describe just about any micro-spherule found on the ground in order to avoid use of the word *micrometeorite*.

But ablation spherules are not true micrometeorites despite their extraterrestrial origin, because they were not small in space. They are more closely related to the meteoritic fusion crust that forms on meteoroids.

The ablation spherules in this book, which range in size from about 0.1 to 0.2 millimeter, are from the meteor that exploded over Chelyabinsk, Russia, on February 15, 2013. They appeared as black dust on the fresh snow, and it is estimated that 12,000 to 13,000 metric tons of the large meteoroid suffered ablation in the atmosphere. The dust plume then unexpectedly streamed back upward into the stratosphere along the jet streams. Within seven days the cloud of ablated particles covered the entire Northern Hemisphere before the spherules eventually fell to the ground.

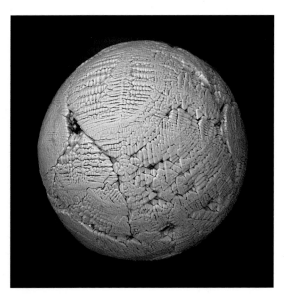

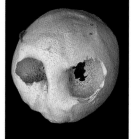

ENIGMATIC CHONDRULES

At 00:30 GMT on October 23, 2012, a fireball was seen over the Izarzar and Beni Yacoub villages near Tata in southern Morocco. The strewnfield was searched extensively, but the meteorite had been extremely friable, and most of it disintegrated midflight. Only small crusted fragments and loose chondrules were found. Some of these (~0.8 to 3.0 millimeter) collected within days of the fall are shown here.

Chondrules are millimeter-sized igneous droplets found in primitive meteorites. They formed in flash-heating events in the Solar Nebula around 4.56 billion years ago (160 million years older than the oldest mineral fragment found on Earth). Most coarse-grained micrometeorites are thought to originate from chondrules.

INDUSTRIAL SPHERULES

MAGNETIC I-TYPE

Magnetic I-type cosmic spherules are iron oxides, mainly magnetite (see page 104) and wüstite. They are resistant to weathering but rare, amounting to only around 2 percent of the overall number of melted micrometeorites found. In deep-sea collections, however, they occur more frequently than 2 percent due to the lower weathering resistance of stony micrometeorites in that environment.

In the search for magnetic micrometeorites in urban areas, we find large numbers of I-type spherules, but these urban I types are not extraterrestrial—their origin is anthropogenic. Many mechanical and industrial processes and all sorts of power tools (oxy-fuel cutting torches, grinding wheels, angle cutters) produce I-type spherules, and they are distributed everywhere by the wind, rain, and human activities.

An energy-dispersive X-ray spectroscopy (EDS) analysis cannot reveal the origin of an I-type spherule unless additional clues, such as nickel beads or platinum group nuggets, can be found. Consequently, the I-type spherules found in populated areas, such as the ones shown here, should be considered anthropogenic unless verified extraterrestrial.

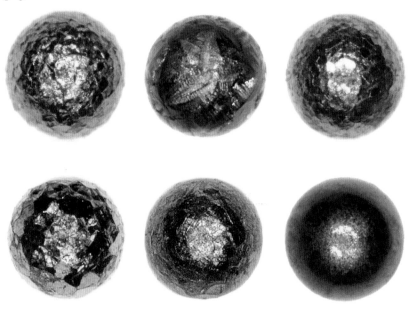

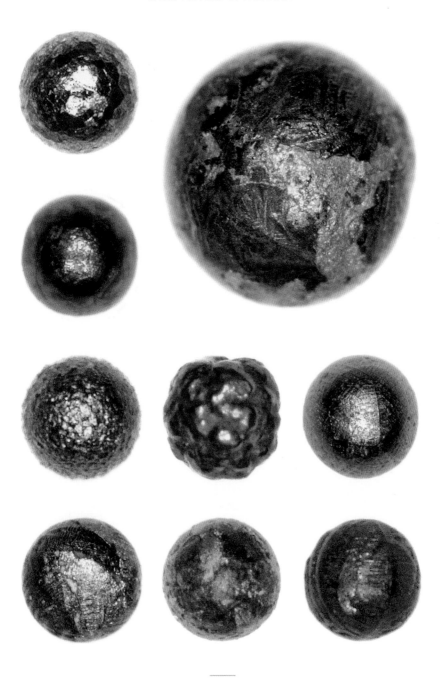

INDUSTRIAL SPHERULES / MASSIVE IRON

MASSIVE IRON

Regular I-type (iron) spherules, both extraterrestrial and anthropogenic, may contain a central void due to rapid solidification inward from the surface. This is not the case with massive I-type spherules, which also have a different morphology than regular I types, though the chemical spectrum is the same: iron oxides. Massive iron spherules are often slightly elongated, sometimes with rudimentary polygonal faces. Unlike regular I-type spherules, massive I types rust quickly, appearing like tiny, rusty cannon balls.

The exact origin of these 0.5- to 1.0-millimeter spherules is uncertain but terrestrial. They show no sign of aerodynamics or hurtling through the atmosphere. The sheer number found along roads, especially on curves and steep hills, may point toward vehicles—perhaps the brake systems of heavy trucks—but this is merely speculation. The point is that massive I-type spherules are not extraterrestrial, but it is necessary to know them well enough to recognize and disregard them when searching for micrometeorites in road dust.

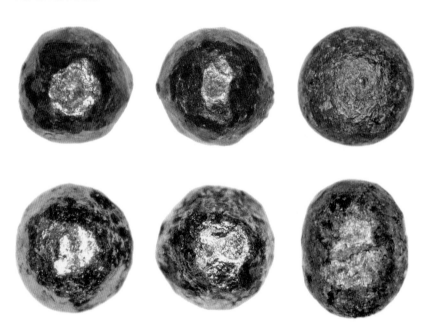

NUGGETS, BEADS & CORES

Platinum group nuggets (PGNs), nickel/iron/chromium-containing beads, and other types of metal inclusions are characteristic features in approximately 5 percent of all cosmic spherules. They vary from submicron-sized PGNs to large nuggets, or "cores." Some micrometeorites have a hole where a metal bead has escaped (see page 56), so it should also be possible to find loose nickel and iron cores.

The differentiation where the heavier elements sink inward and form a core happens rapidly when the particle is in a liquid state, and the surface tension gives the spheroid form. This also happens with anthropogenic spherules such as those shown. Note the lack of aerodynamic forms and absence of magnetite "Christmas trees" compared with the micrometeorites.

The vertical sidebar text on the left margin reads: INDUSTRIAL SPHERULES / NUGGETS, BEADS & CORES

INDUSTRIAL SPHERULES / FROM THE WELDING SHOP

FROM THE WELDING SHOP

The main challenge in the search for micrometeorites in populated areas is distinguishing extraterrestrial particles from terrestrial. Here we can take a lesson from the 2,500-year-old Chinese book *The Art of War*, which advises, "know your enemy." Inspired by this ancient wisdom, I went to a welding shop and asked to sweep the floor.

Subsequent examination of the dust samples under a binocular microscope revealed a panorama of magnetic spherules with a variety of morphological features dominated by composite-particle, twin, splash, and crash formations—which rule out extraterrestrial origin. Although we cannot know exactly which spherule originates from which power tool, we can recognize the anthropogenic "signatures" and narrow the possibly extraterrestrial candidates.

FROM SPARKS

In a not-too-distant past when smoking cigarettes was more common, it was assumed that looking for micrometeorites in populated areas would be impossible because each lighter spark caused by rubbing ferrocerium upon steel when lighting a cigarette created spherules indistinguishable from micrometeorites.

The spherules pictured were created by igniting a lighter so the sparks hit a carbon plate used for SEM examination. Numerous spherules were found, but they do not have chondritic chemistry, and they show traces of rare earth elements, especially cerium (Ce) and lanthanum (La). Note the scale bar and size of the spherules: a maximum of ~5 to 6 microns. This is one-fiftieth the size of an average micrometeorite and one-tenth of the practical lower limit of spherules in micrometeorite collections. By excluding spherules less than 50 microns, we can rule out many contaminants.

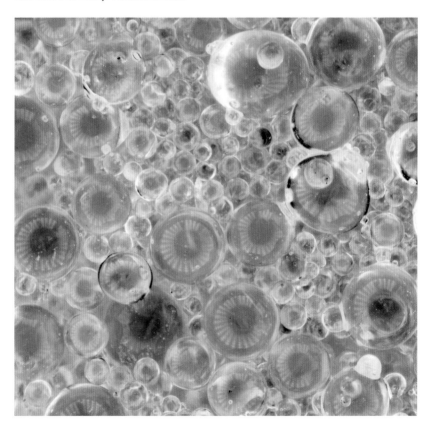

NONMAGNETIC GLASS

In populated areas all over the world, the most common type of all spherules is the omnipresent nonmagnetic glass spherules from road-surface markings. These retroreflective glass beads are produced industrially of flawless, colorless silica glass in defined fractions. Sometimes they constitute the main part of urban road dust.

A wide range of anthropogenic glass spherules are also produced by power tools, industry, and other human activities. These may have colors and vesicles reminiscent of cosmic glass spherules, but their terrestrial origin is often revealed by composite forms, multiple tails, and other clues.

Up to 98 percent of the total number of cosmic spherules on Earth are stony, including the completely melted glass spherules. These V-type (vitreous) micro-meteorites are usually spherical and transparent but can be highly vesiculated. They are colorless to brown or green and normally nonmagnetic unless they contain nickel-iron beads. So far it has not been possible to separate these cosmic glass spherules from the anthropogenic glass spherules described above by mor-phology alone. Chemical analysis is necessary to verify the chondritic spectrum and extraterrestrial origin. Consequently, as with I-type spherules (see page 74), all glass spherules in populated areas can be considered anthropogenic unless verified as extraterrestrial.

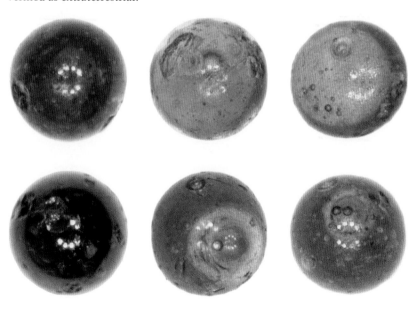

INDUSTRIAL SPHERULES / NONMAGNETIC GLASS

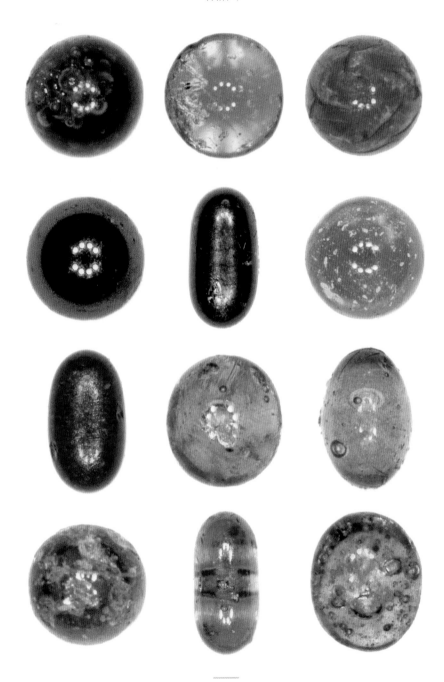

FROM STEAM LOCOMOTIVES

There is a widespread misconception that searching for micrometeorites near railroad lines is impossible due to spherules produced by the trains. In the early days of micrometeorite research, contamination from steam locomotives may have discouraged more than one good scientist, but today these veteran locomotives are rare.

The photos on this page are of micro-objects found in the boiler and the smoke box of an old steam locomotive. The spherules were produced by small mineral impurities in the fuel, in this case high-quality Polish coal. They display an impressive variation in morphology.

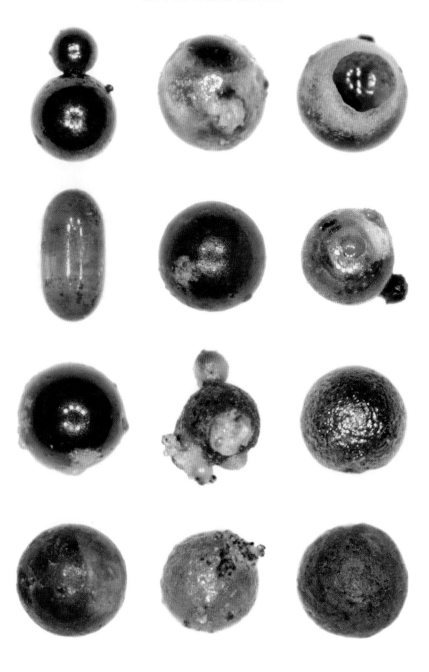

MINERAL WOOL

Mineral wool, such as the Rockwool particles shown, is produced from basalt and chalk. The rocks are heated to a liquid at 1,600°C (2,900°F) and then blown into a large spinning chamber, which pulls the molten rocks into long fibers. At the end of a fiber a droplet may form, and spherules are common, often with one or more tails. The spherules are nonmagnetic and have terrestrial basaltic chemistry.

Mineral wool is used worldwide as insulation, and droplets and spherules such as these are found in the most unexpected places. Simple chemical analysis will reveal whether one is chondritic (a micrometeorite).

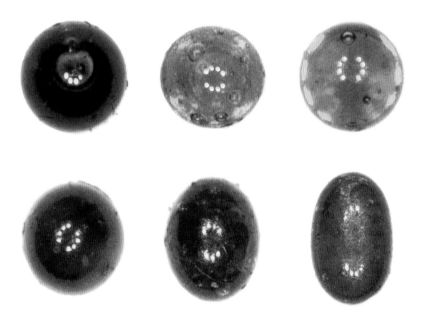

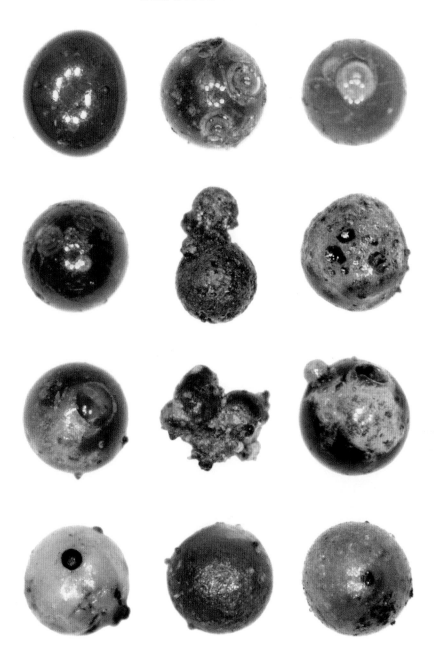

FROM FIREWORKS

Fireworks and flares are chemistry at its most spectacular, but each pyrotechnic spark can create a spherule, and in certain areas these are abundant. Various exotic metal salts (mainly carbonates and chlorides) are used for colorization: strontium/ lithium (red), calcium (orange), sodium (yellow), barium (green), aluminum/tita-nium/magnesium (silver or white), copper (blue), and so on. Flares are often based on strontium nitrate, potassium nitrate, or potassium perchlorate.

Most of the spherules from fireworks can be identified as such under the microscope, either by their vivid colors, by their morphological structures, or by their characteristic metal nugget shape. In most cases a chemical EDS analysis will remove any doubt.

There are, however, some problematic spherules that look like micromete-orites but may actually have been created by fireworks. Some seem to have a near-chondritic chemistry plus barium; others look like turtlebacks (see page 56). Apparently, the sparks from fireworks can go through enough fire and friction to create spherules reminiscent of micrometeorites. They will, however, lack the micrometeoritic Christmas trees. This is one area in need of more research to improve the level of precision in separating micrometeorites in populated areas.

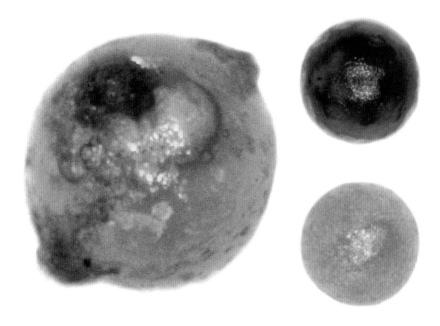

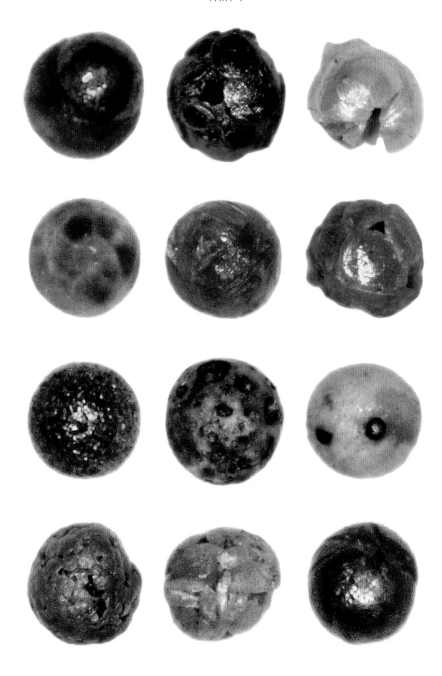

BLACK MAGNETIC (ASPHALT)

There are many references to black magnetic spherules (BMS) in older literature about micrometeorites. The term was often applied to actual micrometeorites. Today in populated areas, however, we encounter a variety of terrestrial spherules that can best be described as BMS. These occur abundantly (a common indication of terrestrial origin). The main sources for BMS are roadway asphalt, roofing shingles, and asphalt glue, which also may contain spherulitic additives, such as aluminum silicates or iron. The morphology ranges from shiny black glass (vitreous or subvitreous) to dull gray via metallic or submetallic with a variety of surface structures, and from perfect spheres to composite forms. Inside, the black magnetic spherules are most often homogenous black, sometimes with a small central vesicle and a radiating structure.

Black carbon (fly ash) is produced by incomplete carbon combustion, both natural and industrial, and these aerosol particles are found worldwide. They are easily distinguished from BMS by their difference in size: black carbon has an average diameter of 0.0025 millimeters (2.5 microns), less than 1 percent of the size of an average micrometeorite.

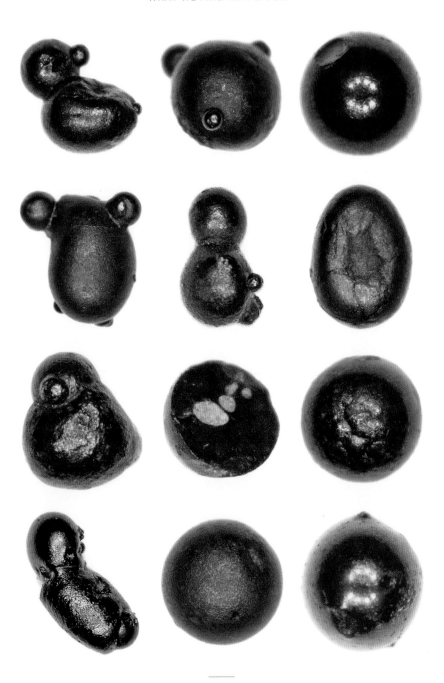

FROM ROOF TILES & SHINGLES

When searching for micrometeorites on roofs, one may encounter various erosion byproducts from the roof itself. Ceramic tiles on a pitched roof are traditionally made from terracotta or concrete and often covered with fine sand in pigmented, iron-rich glue. Over time these magnetic particles (0.2 to 1.0 millimeters) erode and sometimes fill the gutter. For erosion products from asphalt (bitumen) shingles, see page 93.

A roof composed of slate shingle will erode to mainly nonmagnetic mineral particles and provides an excellent hunting ground for micrometeorites.

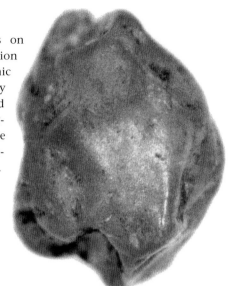

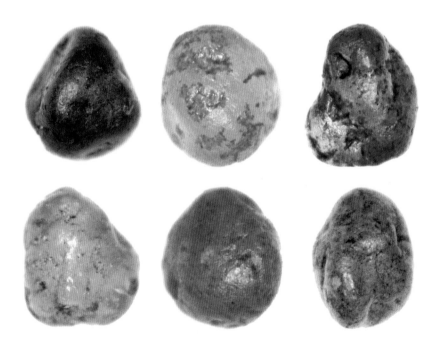

TRACES OF MAN

The industrial revolution marks a major turning point in history; almost every aspect of daily life was influenced in some way, even geological sedimentations. Worldwide from approximately 1760 onward we find anthropogenic or industrial spherules. Consequently, the search for micrometeorites has been a search for uncontaminated nature: prehistoric layers in the Antarctic, remote deserts, ocean floors, and glaciers.

Now that we know extracting extraterrestrial particles in populated areas is possible, we may prefer to go in the opposite direction—looking for large skyward-facing surfaces and learning how to recognize traces of man. Most anthropogenic spherules can be identified as such by their morphological properties (composite forms, abundance, crash and splash forms, tails, rolling marks, and so on). Many manmade spherules are created in fountains of sparks and may still be partly molten when they hit the ground.

On the other hand, micrometeorites are formed in the upper atmosphere and have characteristic aerodynamic forms and surface structures due to their unique histories. They can never hit the ground in a molten state or occur abundantly. Consideration of these two parameters makes it possible in most cases to distinguish one group from the other.

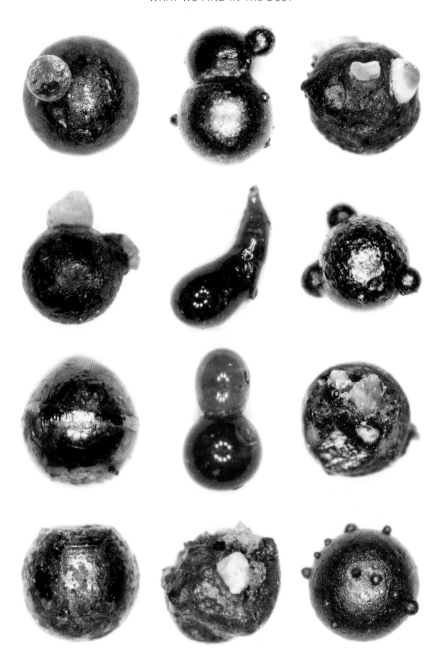

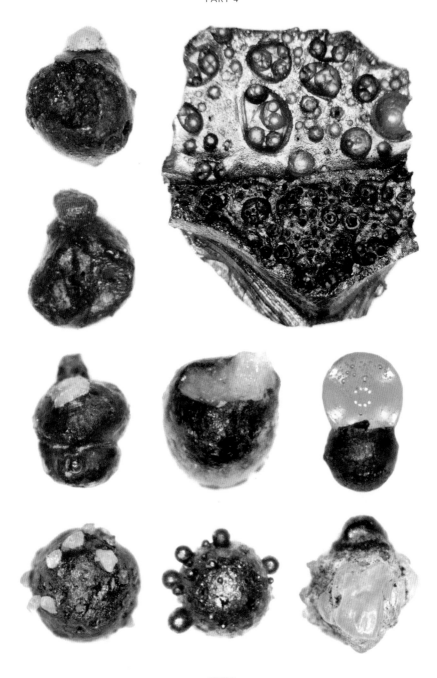

OTHER TYPES

The anthropogenic and naturally occurring terrestrial spherules presented in this book are the most common types and constitute the majority of what there is to be found around the world. The purpose of this presentation is to help facilitate your separation of micrometeorites from samples taken in inhabited areas. There are other types of spherules, but they are rare and/or limited to a local occurrence (where they nonetheless can at times be found in abundance).

There are other types of spheroid concretions, pseudomorphs, microfossils, botryoidal crystallizations, sclerotia, spherulitic mineralization of bacterial origin, industrial spherules from handling all types of materials, and even spherules created by atomic bombs (trinitite).

One of the rapidly expanding types of anthropogenic spherules is microplastics, arguably one of the most characteristic sediment particles of our time. But compared with the micrometeorites in the first part of this book, it should be possible to distinguish most of these types from the extraterrestrial particles by morphology alone or in combination with a simple chemical analysis.

NATURALLY OCCURRING SPHERULES

ROUNDED MINERAL GRAINS

One of the first types of rounded objects we are likely to encounter in the search for cosmic spherules is sand—mineral grains rounded by erosion. Huge amounts of sand are transported around the world by the wind, and total global mineral dust emissions are estimated at 1 billion to 5 billion tons per year, of which 60 million to 200 million tons originate from the Sahara Desert alone. A substantial amount of this is transported across the Atlantic into the Caribbean and Florida.

Consequently, rounded mineral grains will occur in the most unexpected places. A morphological examination under a microscope is usually enough to determine whether the candidate has any of the characteristic structures of a micrometeorite.

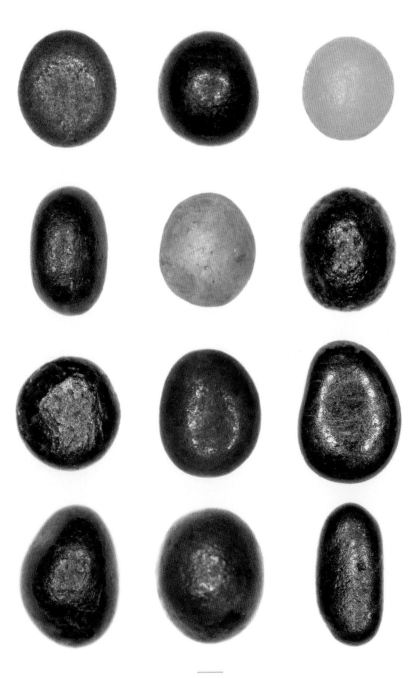

NATURALLY OCCURRING SPHERULES / ROUNDED MINERAL GRAINS

NATURALLY OCCURRING SPHERULES / MAGNETITE

MAGNETITE

Searching for micrometeorites with a magnet is the best path to success. There is, however, one mineral that will occur practically everywhere, and it is also the most magnetic of all naturally occurring minerals on Earth: magnetite. When sampling for micrometeorites in road dust, beach sand, desert sand, mountains, and even on roofs, magnetite particles are abundant and will get caught by the magnet, like the crystals shown here.

FULGURITES

When we search for cosmic spherules we look for melted droplets. These molten micrometeorites distinguish themselves from other mineral grains by their spheroid form and characteristic textures. But in looking for molten mineral grains, we will encounter other, nonmicrometeoritic types.

On Earth, there are three natural processes that melt rock and result in molten mineral grains. The first occurs when desert sand is struck by lightning, resulting in fulgurites, more commonly known as "lightning tubes." A second occurs on rare occasions when a lightning bolt strikes a rock and melts a hole, causing splash droplets, known as exogenic fulgurites. A recently discovered third process happens when lightning strikes the dust plume of an erupting volcano, resulting in molten grains referred to as exogenic volcano fulgurites. (There is a fourth type, exogenic phytofulgurites, possibly the most common of the four, but these are still not properly described.)

The spherules pictured were found at a place where lightning had struck a tree and gone to ground via humus and soil. EDS analysis shows a surprisingly homogenous chemical spectrum despite the wide color range: aluminum silicates (sometimes with an iron oxide rim) and hardly any trace of carbon. This will vary according to the content of the target soil. Approximately one-third of these spherules are magnetic, mainly those with a visible metallic (iron oxide) rim. The spherules are 0.2 to 6.0 millimeter and were found in Ann Arbor and Ypsilanti, Michigan.

On a global scale, there are approximately one hundred lightning strikes per second; obviously the resistant silica spherules these create will not easily dissolve and disappear, so this type of spherule can be expected to be found everywhere.

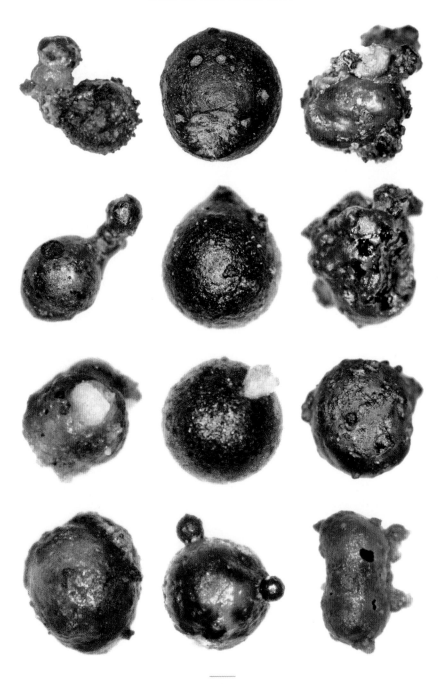

ORGANIC CONFUSION

An interesting stage in the hunt for micrometeorites comes after the field search: visual examination of the dust samples under the microscope. I start with a binocular microscope and pick out promising candidates for further scrutiny under a USB microscope with stronger magnification.

If you search for micrometeorites in dust samples under a microscope, over time you will encounter countless terrestrial particles of all sorts. Sometimes a possible candidate or an unusual object is put aside for thorough inspection, but further investigation may reveal an organic origin: plant seeds, snail shells, insect debris, microfossils, feces, fungi, sclerotia, and so on. Nature is full of beauty, and the photos here show particles that have caused both organic confusion and joy in the hunt for the extraterrestrial.

MICROTEKTITES & MICROKRYSTITES

These are submillimeter-sized spherules formed by large asteroid impacts on Earth. They predominantly comprise melted and vaporized terrestrial target rock. Microtektites, which are wholly glassy by definition, can be derived from individual melt droplets, presenting in the form of ablated macrotektite material and vapor condensates, which form a distinct group. Microkrystites are thought to be vapor condensates and differ from microtektites in that they are part glassy and partly crystalline. Proximal impactite splash droplets are likely to have a degree of crystallinity due to a more basic composition derived from a greater degree of meteoritic contamination of the target rock.

Microtektites are found in defined strata in three of the five established tektite strewnfields on Earth. Spherules can also be found in connection with other, smaller craters, such as the rare S-type spherules from Meteor Crater, Arizona, or the altered spherules from the Bristol impactite layer in England. The most famous microtektites are those found at the iridium-enriched Cretaceous–Paleogene boundary, originating from the great Chicxulub Crater in Yucatán, Mexico. The spherules on the following pages are from this event 66 million years ago. The one at the top was found in Saskatchewan, Canada, and the black ones in Hell Creek Formation, Hardin, Montana.

A rare Darwin glass impactite spherule together with Darwin glass, an impactite found in Tasmania. Despite being created by a prehistoric meteorite impact, these microkrystites are not extraterrestrial but mainly splash droplets of melted and recrystallized terrestrial rock.

NATURALLY OCCURRING SPHERULES / MICROTEKTITES & MICROKRYSTITES

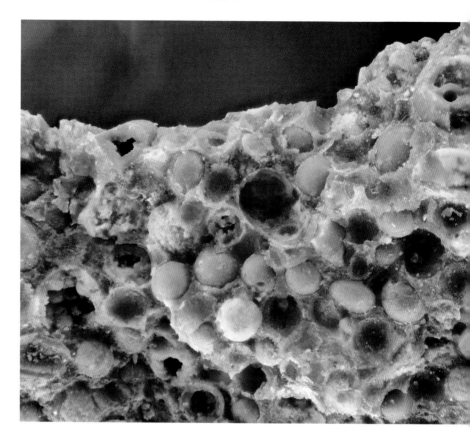

NATURALLY OCCURRING SPHERULES / MICROTEKTITES & MICROKRYSTITES

LONAR CRATER SPHERULES

In the Buldhana district, Maharashtra, India, there is a crater measuring 1.88 kilometers (1.17 miles) in diameter created by a meteorite impact estimated to have occurred more than half a million years ago. Scattered around the well-preserved crater are impactite droplets that differ from most other spherules on Earth and are assumed to be far more common on the Moon and Mars: basaltic impactite melt droplets. Compared with the chemical composition of the surrounding basalt, the spherules are rich in chromium, cobalt, and nickel believed to be remnants of a chondritic meteorite.

The spherules seen here were collected at the Lonar crater rim. They are slightly magnetic and range in size from 0.5 to 14 millimeters.

VOLCHOVITES: A RUSSIAN MYSTERY

The enigmatic volchovites are tektite-like glasses or microkrystites measuring 0.1 to 3.0 millimeters. Mafic and ultramafic in composition (i.e., mainly of ferromagnesium), they were found in the glaciofluvial drift along the Volkhov River near St. Petersburg, Russia. They were discovered and described by Gennady Skublov as cryptovolcanic glass. In a recent study the volchovites have been subdivided into four groups according to the trace elements found in the spherules.

The SEM images here show the microkrystite structure of a volchovite with relatively large crystals around an iron oxide microspherule. The volchovites are reported to occur together with particles of quenched glass, cinder, and rounded fragments of rocks. Some of the spherules contain small metal beads of various contents: titanium, iron, gold, and copper—such as the ones shown below and on the next pages, frozen in time in the very last moment before escaping.

NATURALLY OCCURRING SPHERULES / VOLCHOVITES: A RUSSIAN MYSTERY

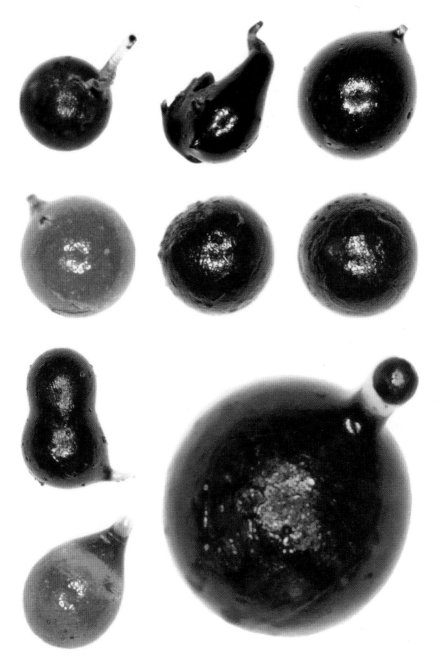

NATURALLY OCCURRING SPHERULES / IBERULITES

IBERULITES

The Iberian Peninsula is regularly sandblasted by particles from Africa. But in between the sand grains (see page 102) are pale spherules of a different kind. They are white to sand-colored, often with a vertical axis, and sometimes with a characteristic vortex; some have a smooth white surface, while others seem to be balls of sand glued together with carbonates. Unlike regular sand grains, they are easily crushed to reveal a coarse-grained core (mineral particles in the 1 to 2 micron fraction) with a thick rind of fine-grained white clay minerals.

Iberulites develop in the troposphere before they fall to Earth's surface. They are linked to the evolution of high-dust air plumes that originate in Saharan dust storms and are transported over the Iberian Peninsula and often across the North Atlantic Ocean. One may assume similar spherules are created in the other grand deserts of the world.

The spherules (~0.3 to 0.9 millimeter) shown were found along the Costa del Sol, Spain.

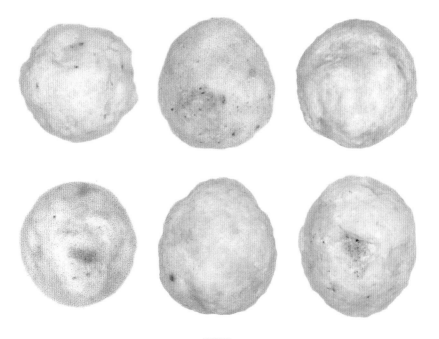

OOIDS & PISOIDS

Because micrometeorites are rare relative to terrestrial and anthropogenic spherules, an abundance of spherules can be a sign of terrestrial origin. Here is a type that can be rock-forming and, as such, should cause little confusion in the hunt for the extraterrestrial. Nevertheless, it is worth being aware of them.

An ooid consists of a nucleus (usually a mineral grain or biogenic fragment) around which concentric layers of minerals are deposited to form a spherical grain of 0.25 to 2 millimeters. Oolith is sediment consisting of ooids. These are most commonly composed of calcium carbonate (the minerals calcite and aragonite) but can also be composed of phosphate, silica (chert), dolomite, or iron minerals.

Ooids usually form in warm, shallow, highly agitated intertidal marine environments, though some form in inland lakes. The mechanism of formation starts with a small fragment acting as a "seed" (nucleus). Strong intertidal currents wash these seeds around on the seabed, where they accumulate layers of chemically precipitated calcite from the supersaturated water. Oolith is commonly found in large current bedding structures resembling sand dunes.

Ooids larger than 2 millimeters are referred to as pisoids. The freshwater pisoids seen here (4 to 5 millimeters) are from the Cretaceous period (about 97 to 112 million years ago) and were found near Taouz, southeast of Erfoud in Morocco.

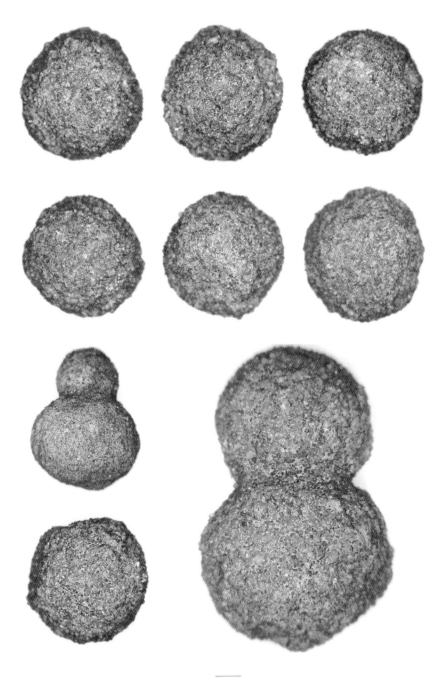

PELE'S TEARS

Among the many types of mineral dust produced and moved around the globe by wind, volcanic tephra is the most dramatic. Entire civilizations have been buried by volcanic eruptions; today, volcanic ash causes problems for air traffic and other modern endeavors. Some volcanic dust particles may, at first glance, look like S-type glassy micrometeorites but can be distinguished from micrometeorites by a simple chemical analysis.

Pele's tears, or *achneliths*, are spherical pyroclasts: drops of volcanic glass thrown out during an eruption. The glassy particles are formed by the quenching of magma spray and are typically 2 to 64 millimeters, or ten to one hundred times larger than average micrometeorites.

Achneliths can reveal a great deal of information about the eruption that created them. Examination of bubbles of gas and particles trapped within the tears can provide information about the composition of the magma chamber, and the shape of the tears can indicate the velocity of the eruption. These achneliths are only 0.3 to 1.5 millimeters and were found on the crater rim of the volcano Le Piton de la Fournaise, Réunion, in the Indian Ocean.

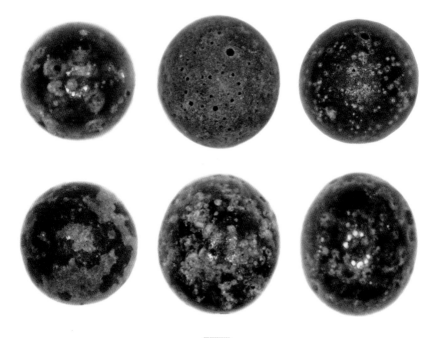

ROAD-DUST CRYSTALS

In the hunt for micrometeorites, we may encounter all sorts of objects while examining samples of road dust under the microscope, occasionally including nature's own jewelry: crystals. Without knowledge of the mineral kingdom one could suspect these amazing geometrical forms, colorful gems, and shiny metal to be not of this world, but well-developed crystals of this size indicate terrestrial origin.

This collection of such submillimeter surprises was found in common road dust while searching for micrometeorites—art by accident. See also the magnetite crystals on page 104. In pursuit of the extraterrestrial, this is yet another category of particles of which to be aware.

5

VERIFY, CLASSIFY, PHOTOGRAPH

MICROMETEORITE VERIFICATION & PHOTOGRAPHY

One of the most common questions about micrometeorites is "How can they be verified?" The short answer is, by determining whether they are *chondritic* and have the right textures.

The definitive evidence for extraterrestrial origin of particles came more than 25 years ago based on noble-gas measurements and analysis of cosmogenic nuclei. All particles exposed to the high-energy cosmic radiation outside Earth's magnetosphere are altered, and these changes in atomic structure can be measured in mass spectrometric analysis.

There are three nonisotopic criteria for the positive identification of a micrometeorite:

First, most micrometeorites have a chondritic chemical bulk composition for major and minor elements (at least for those particles with a small grain size relative to particle size), which is easy to check in an analysis. An example of a typical micrometeorite spectrum is shown on page 130. The three characteristic peaks for the presence of magnesium, silicon, and oxygen are clearly visible, in addition to smaller amounts of aluminum, calcium, and iron.

Second, the presence of nickel-bearing metal in a spherule may suggest extraterrestrial origin, though a lack of nickel does not exclude the possibility, as nickel and iron often differentiate into a core inside a micrometeorite and are not detectable on the surface. In many cases the inertia of the heavy metal pushes the core forward in the direction of fall through the atmosphere during deceleration and ablation, and on nearly 5 to 10 percent of cosmic spherules a metal bead can be seen on the surface, sometimes with an iron-oxide rim covering the nickel inside so that it is not detectable.

The third criterion is the presence of a partial or complete rim of magnetite around the micrometeorite.

The zodiacal cloud is visible right after sunset and before sunrise. This is where micrometeorites of various origin are stored temporarily in the ecliptic plane before they eventually fall into the sun or onto one of the celestial bodies, such as Earth. *ESO, La Silla/Creative Commons*

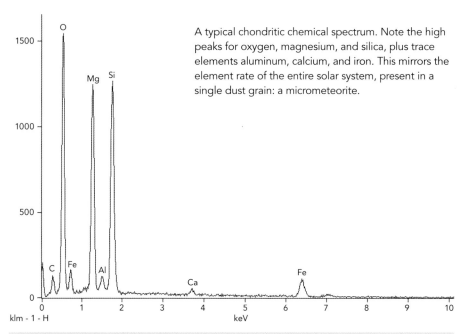

A typical chondritic chemical spectrum. Note the high peaks for oxygen, magnesium, and silica, plus trace elements aluminum, calcium, and iron. This mirrors the element rate of the entire solar system, present in a single dust grain: a micrometeorite.

QUANTITATIVE RESULTS   BASE (36)

Element Line	Weight %	Formula	Compound %
C K	16.02	C	16.02
O K	47.31	O	47.31
Mg K	11.21	Mg	11.21
Al K	0.80	Al	0.80
Si K	9.60	Si	9.60
Ca K	0.65	Ca	0.65
Fe K	14.41	Fe	14.41
TOTAL	100.00		100.00

In addition to these three positive IDs for micrometeorites, there are supporting but less definitive features, such as olivines high in calcium and chromium and very low in iron, which are exceedingly rare in terrestrial rocks.

The discovery and study of urban micrometeorites have shown that in most cases, it is possible to identify fresh micrometeorites on their morphology alone. The unique process micrometeorites have gone through during atmospheric flight—melting, differentiation, and recrystallization at hyperspeed—causes distinct structures not present in terrestrial mineral particles. With some experience it is possible to recognize these and the characteristic aerodynamic forms of the micrometeorites. In combination with knowledge of the most common types of anthropogenic and naturally occurring spherules it becomes possible for anyone to pick out micrometeorites in dust samples from populated areas.

ORIGIN, FORMATION, INFLUX & CLASSIFICATION

There are almost as many explanations as to where micrometeorites originate as there are researchers in the field. Depending upon whom you ask, the answer may vary from the asteroid belt between Mars and Jupiter, comet-related objects in the Kuiper belt or Oort cloud, various planetary ejecta, interstellar matter, and so on. It is estimated that up to 0.1 percent of the matter in primitive meteorites, and possibly also in micrometeorites, comprises presolar grains.

On the other hand, there are *achondritic* (igneous) micrometeorites from differentiated bodies, such as the Moon and the Vesta asteroid. Throughout history, large asteroid impacts on the rocky planets and their moons have ejected substantial quantities of rocks into space, and it is possible to imagine an extensive exchange of matter among all the planetary bodies, their geysers, and surrounding dust rings, with the zodiacal cloud as a temporary storage pool. Such a theory was considered sci-fi only a few years ago.

Hopefully potentially large sources of micrometeorites in populated areas will contribute to a systematic mapping of the isotopic variations of a substantial number of micrometeorites, resulting in more data about the micrometeorites' parent bodies. It should not come as a surprise if the origins of micrometeorites turn out to be a combination of *all* dust-producing bodies in the solar system and beyond. Every micrometeorite hunt has the possibility to sample small rocks from the entire cosmic dust complex.

Micrometeoroids enter Earth's atmosphere with speeds up to fifty times that of a rifle bullet. Depending upon their entry angle relative to Earth's rotation, their peak temperature from the frictional heat will cause a substantial variation in the alteration process. Approximately half of the micrometeoroids in the <0.1-millimeter fraction receive a soft deceleration and end on the ground as unmelted micrometeorites. The rest reach peak temperatures between 1,350 and 2,000°C (2,500 to 3,600°F), enough to create the various types of melted cosmic spherules (see page 50).

At the same time, a rapid differentiation takes place, by which the heavier elements (iron, nickel, platinum) move inward to form a core and volatile elements escape. Iron from the stone reacts with oxygen in the atmosphere and creates

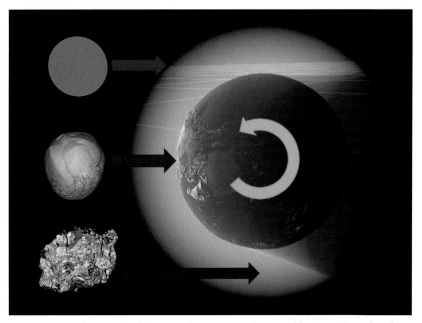

Micrometeoroids enter Earth's atmosphere at speeds up to fifty times that of a rifle bullet. Depending upon their entry angle relative to Earth's rotation, their peak temperature from frictional heat causes substantial variation in the alteration process and helps determine what kind of micrometeorite is formed.

dendritic magnetite, which looks like small Christmas trees on the surface. With a micrometeroid still in flight but decelerating, the inertia of its heavy core may push it forward in the direction of travel, often in spin. The whole formation is over in the blink of an eye before the micrometeoroid falls to Earth at terminal velocity. Based on radar measurements, the general influx rate of micrometeorites is estimated to be approximately one object with a diameter of 0.1 millimeter per square meter per year.

"The Classification of Micrometeorites," published in 2008 by Matthew Genge with Cécile Engrand, Matthieu Gounelle, and Susan Taylor, is the most comprehensive article about micrometeorites to date. It is freely available on the Internet and is a must for anyone interested in the subject. The authors define the various types of micrometeorites by their characteristic textures, which are mainly determined by the peak temperature they reach in combination with the quenching profile (fast or slow cooling) during atmospheric flight. As mentioned previously, there are transitional forms between the various types, but the

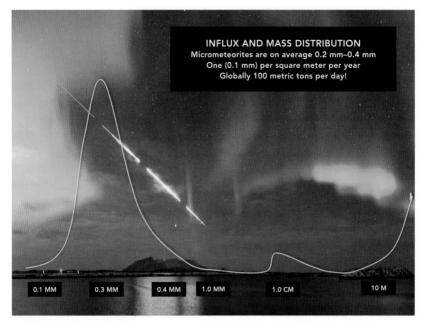

INFLUX AND MASS DISTRIBUTION
Micrometeorites are on average 0.2 mm–0.4 mm
One (0.1 mm) per square meter per year
Globally 100 metric tons per day!

| 0.1 MM | 0.3 MM | 0.4 MM | 1.0 MM | 1.0 CM | 10 M |

The mass distribution of micrometeorites. The x-axis is increasingly condensed toward the right but shows the large peak of the cosmic dust particles between 200 and 400µm. From around 1cm we have the meteorites, substantially fewer than the micrometeorites. At the far right we find the rare but devastating asteroids.

chemistry of micrometeorites is surprisingly homogenous—mainly chondritic (see spectrum on page 130) with some minor (or rare) variations. Future research may add varieties to the present classification, and with more hands and eyes in the field, micrometeoritics can evolve into an exciting new branch of the popular study of space rocks.

The mass distribution of the extraterrestrial influx on Earth is shown on page 130. The size scale is increasingly compressed to the right on the x-axis. To the left are the micrometeorites, with a distinct peak between 0.2 and 0.4 millimeters, before it decreases to 0 between approximately 2 millimeters and 1 centimeter, which is where meteors appear. With larger mass and kinetic energy, meteors burn up in the atmosphere, leaving behind only nanosized meteoritic smoke particles. From around 1 centimeter upward to a few meters are the meteorites, and to the right are the large, but rare, asteroids (large bodies in orbit around the sun).

Micrometeorites are the smallest of the three groups of extraterrestrial matter, with an influx rate of one object with a diameter of approximately 0.1 millimeter per square meter per year. The average cosmic spherule, however, has a diameter around 0.3 millimeter, which contains up to twenty-seven times more matter than one at 0.1 millimeter. A large micrometeorite with a diameter 0.45 millimeter contains no less than ninety-one times more mass than one at 0.1 millimeter. Consequently, in our hunt for micrometeorites in populated areas, on a roof of 100 square meters, we cannot expect to find one hundred cosmic spherules per year but one to four.

As mentioned, micrometeoroids enter Earth's atmosphere at speeds of up to fifty times that of a rifle bullet, and, depending upon their entry angle relative to Earth's rotation, the peak temperature from the frictional heat will cause a substantial variation in the formation process. In the illustration on page 132 we see a completely melted (glassy) micrometeorite on top, a fine-grained/barred olivine micrometeorite in the middle, and an unmelted micrometeorite on the bottom.

PHOTOGRAPHING MICROMETEORITES

The traditional way to depict micrometeorites has been via scanning electron microscope (SEM) images. In a backscatter SEM image, the heavy elements, such as iron and nickel, are a lighter (whiter) contrast to the darker stone. To *see* the chemical distribution like this is usually sufficient to identify whether an object is terrestrial or extraterrestrial. In addition, SEM images can show extremely high resolution and magnification. What these images do not show, however, are the natural colors and transparence or translucence. A green glass spherule with a golden metal core inside will appear as a white sphere in the SEM images.

Color photography is therefore a useful addition to the traditional way of documenting micrometeorites and a popular way to show your findings. It can be done in several ways, and with today's rapid development in digital photographic equipment, new methods are introduced every year.

Possibly the easiest way to photograph micrometeorites is with a USB microscope and a magnification around 200×, with the pictures saved directly to a hard drive. With existing technology these photos are not in very high resolution, but this will continue to improve.

Many collectors even use their smartphone to photograph through the microscope and manage to make stunning micrometeorite photos this way. Special microscope lenses are available for smartphones, and they are getting better all the time.

The photo rack constructed by Jan Braly Kihle and the author to take high-resolution color photos of micrometeorites. The camera moves up and down at 1μm steps, and the micrometeorite is illuminated by a specially invented rib-cooled ring lamp. This photo instrument can take up to 3,000×-magnification photos of space dust, but the process is time consuming and takes two people to operate. In one day, we can finish only two new color photos.

A trinocular microscope is a binocular with a third ocular for a camera. These usually make excellent photos, as the equipment is specially made for the purpose. But like all special equipment, they are expensive.

The most important part of micrometeorite photography is to document your findings and share the pictures with others. We are pioneers in a new scientific branch and have an obligation to share this mind-blowing jewelry from space with others.

A returning challenge with microphotography is the lack of depth of field at higher magnifications. This can be solved with stacking technology: software developed specifically to take a succession of individual exposures gradually moving the focal point up or down between exposures until the entire surface of the object has been in focus. The exposures are then "stacked" in a single photo. Many stacking programs are available, and some new cameras even have the ability integrated into their software. Color correction and the removal of noise in the pictures and even unwanted terrestrial dust on the subject is done afterwards.

CAMERA LAYOUT

1. µ4/3 camera
2. Optics
3. LED ring with heat-dissipating aluminum fins
4. Flexible silver wire ring light holder
5. Ring-light height-adjustment screws
6. X-Y micro positioner
7. Covered glass sample holder
8. Sorbothane vibration insulators (three densities)
9. Solid aluminum block
10. Stepper bracket
11. Z-axis rail
12. Piezo motor for Z-axis rail
13. Bracket for camera
14. Vacuum tweezer with holder for lifting glass cover
15. Power supplies, controllers, and PC interfaces
16. 20mm optical steel table
17. A4 steel bolts

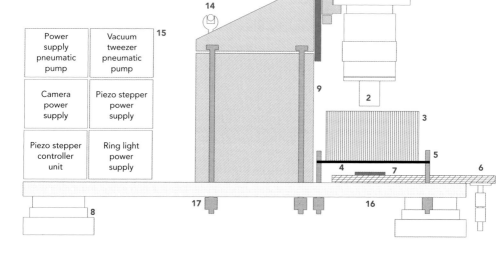

To take extremely high-resolution color photos of micrometeorites, Jan Braly Kihle and I have constructed a special instrument, shown on page 135, based on an Olympus EM1 Mark II camera (50MB) in combination with an all-metal telescopic tubus and ultralong working distance microscope objectives on a motorized rack allowing the camera to be moved up and down with high precision, one micron step at a time. For the highest-resolution imaging of sub-150-micron particles, a slightly modified setup is applied using an apochro-

This photo gives good perspective on the size of micrometeorites. Sixteen micro-meteorites are mounted on a white sticky note glued to glass and ready for photography. The author's and Kihle's photo rack is visible in the background.

matic supertelephoto lens in combination with an infinity-corrected microscope objective. The micrometeorite is mounted on a white sticky note on glass laid on a micropositioning X-Y table lighted with a specially constructed rib-cooled LED ring light. All data from the camera is captured by computer in real time. When the sequence of single exposures (the average is 150 to 200 individual pictures per micrometeorite) is completed, the stacking program assembles them into one color photo with the entire hemisphere of the stone in focus. All the color photos of micrometeorites in this book were made with this method.

Developing the components, inventing the various modifications, writing the software. and constructing a rig like this has taken many years and cost a lot of money. The reward is that we can see the micrometeorites like nobody else has before, which has changed the way we think about cosmic dust particles. They are not dust, but small individual stones of amazing beauty, telling us stories about how it all began, but still guarding secrets we are eager to understand. This exciting new connection with space is now within reach for all of us. Follow the instructions in this book, and welcome to the hunt for stardust.

FURTHER READING

Folco. L., and C. Cordier. 2015. "Micrometeorites." *European Mineralogical Union Notes in Planetary Mineralogy* 15:253–297. DOI: 10.1180/EMU-notes 15.9.

Genge, Matthew J., Cécile Engrand, Matthieu Gounelle, and Susan Taylor. 2008. "The Classification of Micrometeorites." *Meteoritics & Planetary Science* 43, no. 3, 497–515.

Genge, Matthew J., Jon Larsen, Matthias Van Ginneken, and Maartijn D. Suttle. 2017. "An Urban Collection of Modern-Day Large Micrometeorites." *Geology* 45 (2): 119–122.

Larsen, Jon. 2017. *In Search of Stardust.* Voyageur Press. Minneapolis, Minnesota. Originally published in 2016 by Kunstbokforlaget DGB. Also available in Japanese, German, and Chinese.

Larsen, Jon. 2018. *Stjernejeger (Star Hunter).* CappelenDamm, Oslo, Norway.

Larsen, Jon, Morten Bilet, Øivind Thoresen, and Rune Selbekk. 2014. *Norske Meteoritter.* Kunstbokforlaget DGB, Oslo, Norway.

Maurette, Michel. 1993. *Chasseurs d'Etoiles (Star Hunters).* Hachette, Paris, France.

Maurette, Michel. 2006. *Micrometeorites and the Mysteries of Our Origins.* Springer-Verlag, Berlin, Germany.

Peucker-Ehrenbrink, B., and B. Schmitz, editors. 2001. *Accretion of Extraterrestrial Matter throughout Earth's History.* Kluwer Academic/Plenum Publishers, New York.

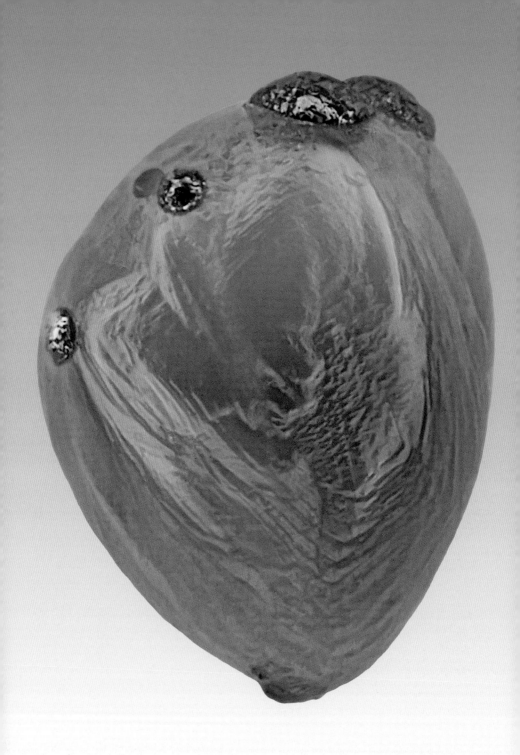

INDEX

ABOUT THE AUTHOR

Jon Larsen (born 1959) is a citizen scientist and jazz guitarist, founder of the group Hot Club de Norvège. He has produced more than 600 jazz records, with legendary musicians such as Chet Baker, Stéphane Grappelli, Warne Marsh, Jimmy Rosenberg, etc. He has received several awards for his musical work and has published several books.

Ever curious, Jon Larsen started his search for micrometeorites in 2009. His breakthrough came in 2015 with the verification of the world's first micrometeorite discovered in a populated area. In 2016, Dr. Matthew Genge at the Natural History Museum in London evaluated and verified Jon Larsen's collection of urban micrometeorites.

Since then, Jon Larsen has been holding lectures all over the world. NASA has even invited him to share his findings and to learn more about micrometeorites. Today he is doing scientific research on micrometeorites at the University of Oslo, Norway.